Advances in Intelligent Systems and Computing

Volume 893

Series editor

Janusz Kacprzyk, Systems Research Institute, Polish Academy of Sciences,
Warsaw, Poland
e-mail: kacprzyk@ibspan.waw.pl

The series "Advances in Intelligent Systems and Computing" contains publications on theory, applications, and design methods of Intelligent Systems and Intelligent Computing. Virtually all disciplines such as engineering, natural sciences, computer and information science, ICT, economics, business, e-commerce, environment, healthcare, life science are covered. The list of topics spans all the areas of modern intelligent systems and computing such as: computational intelligence, soft computing including neural networks, fuzzy systems, evolutionary computing and the fusion of these paradigms, social intelligence, ambient intelligence, computational neuroscience, artificial life, virtual worlds and society, cognitive science and systems, Perception and Vision, DNA and immune based systems, self-organizing and adaptive systems, e-Learning and teaching, human-centered and human-centric computing, recommender systems, intelligent control, robotics and mechatronics including human-machine teaming, knowledge-based paradigms, learning paradigms, machine ethics, intelligent data analysis, knowledge management, intelligent agents, intelligent decision making and support, intelligent network security, trust management, interactive entertainment, Web intelligence and multimedia.

The publications within "Advances in Intelligent Systems and Computing" are primarily proceedings of important conferences, symposia and congresses. They cover significant recent developments in the field, both of a foundational and applicable character. An important characteristic feature of the series is the short publication time and world-wide distribution. This permits a rapid and broad dissemination of research results.

Advisory Board

More information about this series at http://www.springer.com/series/11156

Juan Carlos Corrales · Plamen Angelov
José Antonio Iglesias
Editors

Advances in Information and Communication Technologies for Adapting Agriculture to Climate Change II

Proceedings of the 2nd International Conference of ICT for Adapting Agriculture to Climate Change (AACC'18), November 21–23, 2018, Cali, Colombia

 Springer

Editors
Juan Carlos Corrales
Telematics Department
University of Cauca
Popayán, Colombia

José Antonio Iglesias
Computer Science Department
Carlos III University of Madrid
Leganés, Madrid, Spain

Plamen Angelov
School of Computing and Communications
Lancaster University
Lancaster, UK

ISSN 2194-5357 ISSN 2194-5365 (electronic)
Advances in Intelligent Systems and Computing
ISBN 978-3-030-04446-6 ISBN 978-3-030-04447-3 (eBook)
https://doi.org/10.1007/978-3-030-04447-3

Library of Congress Control Number: 2018961440

This Springer imprint is published by the registered company Springer Nature Switzerland AG
The registered company address is: Gewerbestrasse 11, 6330 Cham, Switzerland

Preface

Over the last decade, information and communications technologies (ICTs) have emerged as tools supporting farming and agricultural production. This is reflected in the multiple hardware and software applications providing services to reduce effects caused by diseases and pests in crops, or in real-time monitoring of weather conditions and water quality parameters used for farming and production. Sustainable supply, risk, and uncertainty management in agriculture productivity increase and cost reductions are some issues that big data solutions and data analytics can help to solve.

This international conference looked at emerging problems and new ICT solutions to address the effects of climate change and climate variability in the agricultural sector, proposing common strategies and guidelines for incorporating risk management and adaptation to climate change. Topics such as smart farming, big data and IoT in agriculture, climate change, and smart irrigation in agriculture were covered by the authors.

The conference was organized by the Inter-Institutional Network for Climate Change and Security Food of Colombia—RICCLISA, University Icesi, University of Cauca (Telematics Engineering Group) Colombia, and the Carlos III University of Madrid, Spain (CAOS research group), and technically sponsored by Springer.

The conference received 24 submissions from 105 authors from 8 countries (Chile, Colombia, India, Spain, Ecuador, Italy, France, USA). This volume collects 17 papers accepted and presented, confirm thus ascertaining its status of the international event. The papers were grouped into the following sessions: smart farming, big data in agriculture, application of IoT in agriculture, ICT to support the agricultural supply chain, climate change, risk and impact analyses on health by water pollution caused by agriculture, and smart irrigation in agriculture.

This conference provided a friendly atmosphere and will be a leading international forum focusing on discussing problems, research, results, and future directions in the application of information and communications technology solutions that allow improving the agricultural production in climate change scenarios.

Finally, we would like to thank the hard work and dedication of the Programme Committee members and Organizing Committee members. Thank you.

September 2018

Juan Carlos Corrales
Jose Antonio Iglesias
Plamen Angelov
AACC18 General Chairs

Organization

AACC 2018 was organized by Red Insterinstitucional de Cambio Climático y Seguridad Alimentaria de Colombia—RICCLISA, Universidad Icesi—Colombia, Universidad del Cauca—Colombia, Universidad Carlos III de Madrid—España, and Centro Regional de Productividad e Innovación del Cauca CREPIC—Colombia.

General Chairs

Juan Carlos Corrales	Universidad del Cauca, Colombia
José Antonio Iglesias	Universidad Carlos III de Madrid, Spain
Plamen Angelov	Lancaster University, UK

Program Co-chairs

Álvaro Pachón	Universidad Icesi, Colombia
Gonzalo Llano Ramírez	Universidad Icesi, Colombia
Juan Carlos Cuellar	Universidad Icesi, Colombia
Emmanuel Lasso	Universidad del Cauca, Colombia
Rafael Bermudez	CREPIC, Colombia

Organizing Committee

Álvaro Pachón	Universidad Icesi, Colombia
Gonzalo Llano Ramirez	Universidad Icesi, Colombia
Andrés López	Universidad Icesi, Colombia
Luis Eduardo Munera	Universidad Icesi, Colombia
Lina Marcela Quintero	Universidad Icesi, Colombia
Juan Carlos Cuellar	Universidad Icesi, Colombia

Program Committee

Wuletawu Abera	CIAT, Colombia
Plamen Angelov	Lancaster University, UK
Sandro Bimonte	IRSTEA, France
Javier Borondo	Universidad Politécnica de Madrid, Spain
Oscar Caicedo	Universidad del Cauca, Colombia
Liseth Viviana Campo Arcos	University of Cauca, Colombia
David Camilo Corrales	Universidad Carlos III de Madrid, Spain
Juan Carlos Corrales	Universidad del Cauca, Colombia
Juan Carlos Cuellar Q.	Universidad Icesi, Colombia
Apolinar Figueroa	Universidad del Cauca, Colombia
Cristhian Figueroa	Politecnico di Torino, Italy
Cesar Garcia	Cenicaña, Colombia
Jorge Gomez	Universidad del Sinú, Colombia
Laura Gonzáles	University of the Republic, Uruguay
German Gutierrez	Universidad Carlos III de Madrid, Spain
Jose Iglesias	Universidad Carlos III de Madrid, Spain
Emmanuel Lasso	Universidad del Cauca, Colombia
Agapito Ledezma	Universidad Carlos III de Madrid, Spain
Gonzalo Llano Ramirez	Universidad Icesi, Colombia
Ivan Dario Lopez	Universidad del Cauca, Colombia
Andrés López	Universidad Icesi, Colombia
Diego López	Universidad del Cauca, Colombia
Carlos Meira	Embrapa Informática Agropecuária, Brazil
Christian Mendoza	Cenicaña, Colombia
Juan Mendoza	USTA Tunja, Colombia
Jordi Morato	Universitat Politècnica de Catalunya, Spain
Luis Munera	Universidad Icesi, Colombia
Hugo Armando Ordoñez Erazo	Universidad de San Buenaventura—Cali, Colombia
Armando Ordóñez	University of Cauca, Colombia
Alvaro Pachon	Universidad Icesi, Colombia
Leonairo Pencue	Universidad del Cauca, Colombia
Andrés Peña	Cenicaña, Colombia
Francois Pinet	Cemagref, France
Julián Eduardo Plazas Pemberthy	Universidad del Cauca, Colombia
Gustavo Ramírez	Universidad del Cauca, Colombia
Alvaro Rendon	Universidad del Cauca, Colombia
Juan Ruiz-Rosero	Universidad del Cauca, Colombia
Araceli Sanchis	Universidad Carlos III de Madrid, Spain
Luz Santamaría	Universidad Santo Tomás Seccional Tunja, Colombia

M. Paz Sesmero	Universidad Carlos III de Madrid, Spain
Tatiana Solano	University of Trento, Italy
Fernando Urbano	Universidad del Cauca, Colombia
Cristian Valencia	Universidad del Cauca, Colombia
Monica Valencia	Universidad del Cauca, Colombia
Héctor Villada	Universidad del Cauca, Colombia

Additional Reviewers

Burbano, Marlon Felipe
Chantre, Angela
Figueroa Milena, Dayana
Ruiz-Rosero, Juan

Sponsors

Red Interinstitucional de
Cambio Climático y Seguridad
Alimentaria (RICCLISA)

Universidad del Cauca

Grupo de Ingeniería
Telemática—Universidad
del Cauca

**Grupo de
Ingeniería Telemática**

Universidad Icesi

Crepic

Universidad Carlos III de
Madrid

Contents

Inference System for Irrigation Scheduling with
an Intelligent Agent 1
Andres Fernando Jimenez, Eric F. Herrera, Brenda V. Ortiz, Antonio Ruiz,
and Pedro Fabian Cardenas

Decision Support System for Precision Irrigation Using Interactive
Maps and Multi-agent Concepts 21
Giovanny Hernández, Andres Fernando Jimenez, Brenda V. Ortiz,
Alfonso P. Lamadrid, and Pedro Fabian Cardenas

IoT Architecture Based on Wireless Sensor Network Applied
to Agricultural Monitoring: A Case of Study of Cacao
Crops in Ecuador 42
Juan Carlos Guillermo, Andrea García-Cedeño, David Rivas-Lalaleo,
Mónica Huerta, and Roger Clotet

Silkworm Growth Monitoring in Second Stage -Instar- Using Artificial
Vision Techniques 58
Luis Javier Suárez, Yaneth Patricia López, Wilfred Fabian Rivera,
and Agapito Ledezma

Affectations in Soil Fertility in the Direct Influence Area Downstream
of the Grande River: Chone Multiple Purpose Dam in Ecuador 73
Oswaldo Borja-Goyes, David Carrera-Villacrés, David González-Riera,
Edgar Guerrón-Varela, Paula Montalvo-Alvarado,
and Andrés Moreno-Chauca

Methodology for Microgrid/Smart Farm Systems: Case of Study
Applied to Indigenous Mapuche Communities 89
Carolina Vargas, Raúl Morales, Doris Sáez, Roberto Hernández,
Carlos Muñoz, Juan Huircán, Enrique Espina, Claudio Alarcón,
Victor Caquilpan, Necul Painemal, Tomislav Roje,
and Roberto Cárdenas

**Dynamics of the Indices NDVI and GNDVI in a Rice Growing
in Its Reproduction Phase from Multi-spectral Aerial Images
Taken by Drones** 106
Diego Alejandro García Cárdenas, Jacipt Alexander Ramón Valencia,
Diego Fernando Alzate Velásquez, and Jordi Rafael Palacios Gonzalez

**Wireless Sensor Network for Monitoring Climatic Variables
and Greenhouse Gases in a Sugarcane Crop** 120
Oscar A. Orozco and Gonzalo Llano

**A Proposed Unmanned and Secured Nursery System
for Photoperiodic Plants with Automatic Irrigation Facility** 136
Fatima Jannat, Tasmiah Tamzid Anannya, Tanu Dewan, Saad Bin Bashar,
Ayeasha Akhter, Ismat Tarik, and Farhana Afroz

**Design and Development of an Intelligent Seed Germination System
Based on IoT** 146
Muhammad Nazrul Islam, Mahmuda Rawnak Jahan, Abid Ali,
Shamsuzzaman Rony, Tasmiah Tamzid Anannya, Faisal Ibn Aziz,
Moin Bayzed, Anika Yeazdani, and Md. Fazle Rabbi

**A Modeling Infrastructure Based on SWAT for the Assessment
of Water and Soil Resources** 162
Pierluigi Cau and Giovanni De Giudici

**Electronic Crop (e-Crop): An Intelligent IoT Solution for Optimum
Crop Production** 177
V. S. Santhosh Mithra

**A Method to Improve the Performance of Raster Selection Based
on a User-Defined Condition: An Example of Application for Agri-
environmental Data** 190
Driss En-Nejjary, François Pinet, and Myoung-Ah Kang

**Coffee Crops Variables Monitoring: A Case of Study
in Ecuadorian Andes** 202
Juan Abad, Juan Farez, Paúl Chasi, Juan Carlos Guillermo,
Andrea García-Cedeño, Roger Clotet, and Mónica Huerta

**Metamorphosis and Dynamics of Inorganic Species in the Waters
of the Chone Multiple Purpose Dam, Ecuador** 218
Rosa Arias-Carrera, Paola Álvarez-Castillo, Oswaldo Borja-Goyes,
Leidy Cajas-Morales, David Carrera-Villacrés, Karla Enríquez-Herrera,
Tania González-Farinango, David González-Riera, Erika Guamán-Pineda,
Solange Guitiérrez-Puetate, Andrés Moreno-Chauca,
Paula Montalvo-Alvarado, Richard Rubio-Gallardo,
Tatiana Sandoval-Plaza, Nelson Unda-León, and Paul Velarde-Salazar

**Evaluation of Threats to Agriculture in the Totaré River Basin Due
to Changes in Rainfall Patterns Under Climate Change Scenarios**..... 234
Freddy Duarte, Jordi Rafael Palacios, and Germán Ricardo Santos

**IoT Network Applied to Agriculture: Monitoring Stations
for Irrigation Management in Soils Cultivated with Sugarcane** 249
Edgar Hincapié Gómez, Juliana Sánchez Benítez,
and Javier Alí Carbonell González

Author Index... 261

Inference System for Irrigation Scheduling with an Intelligent Agent

Andres Fernando Jimenez[1,2,3(✉)], Eric F. Herrera[1], Brenda V. Ortiz[3],
Antonio Ruiz[4], and Pedro Fabian Cardenas[1]

[1] Unrobot Research Group, Universidad Nacional de Colombia, Bogotá, Colombia
{efherreral,pfcardenash}@unal.edu.co
[2] Macrypt Research Group, Universidad de los Llanos, Villavicencio, Meta, Colombia
ajimenez@unillanos.edu.co
[3] Crop, Soil and Environmental Sciences Department, Precision Agriculture,
Auburn University, Auburn, AL, USA
bvo0001@auburn.edu
[4] Department of Engineering, Universidad Miguel Hernández de Elche (UMH),
Orihuela, Spain
aruizcanales@gmail.com

Abstract. The design and implementation of an inference method to schedule irrigation tasks in agriculture, based on the concept of rational agent is explained in this paper. Through the use of a raspberry-pi and xbee devices, the agent interacts with its environment acquiring soil moisture, soil temperature, luminosity, air temperature and rain data. Membership functions and a Mamdani inference methodology were implemented on the raspberry-pi to define irrigation time. Furthermore solenoid valves were used as actuators to apply the amount of water prescribed on the field. The inference system uses luminosity and ambient temperature data to define periods with high levels of evapotranspiration and soil moisture sensors to determine the Volumetric Water Content (VWC). As result of this implementation, temperature membership functions were defined, including the annual averages of minimum ($9\,^{\circ}\mathrm{C}$) and maximum ($20\,^{\circ}\mathrm{C}$) temperatures, according with weather data obtained in Nobsa, Boyacá, Colombia. Values between 0 and 2500 lx, were defined in the universe of discourse of brightness and soil moisture membership was defined using field capacity and permanent wilting points. The system developed allows to maintain the VWC in the soil near to the field capacity value, wherewith, soil moisture is maintained at the optimum level necessary for the correct development of a crop.

Keywords: Artificial intelligence · Fuzzy logic · Irrigation
Precision agriculture

© Springer Nature Switzerland AG 2019
J. C. Corrales et al. (Eds.): AACC 2018, AISC 893, pp. 1–20, 2019.
https://doi.org/10.1007/978-3-030-04447-3_1

1   Introduction

Precision agriculture (PA) is defined as a farm technology based on information, designed to apply agricultural inputs in a controlled manner and site-specific managed. Given that info-based PA approaches, current technologies like sensors data acquisition and processing systems, wireless networking modules, and techniques like data mining as well as machine learning, enhance the management of large amount of data and are powerful tools for research [1]. In precision agriculture, spatial and temporal variability are considered, and is commonly employed by sowing, fertilization, pest control, weed and disease management, harvesting and irrigation, among others.

Irrigation is the process of applying water on the soil, what is useful to improve the growth of agricultural crops, maintain landscapes and revegetate degraded soils in dry areas with irregular periods of rain [2]. However, only a portion of dispensed water is absorbed by the plant. The rest is used to wash salts from the soil or is lost unproductively by evaporation in irrigation channels as well as by runoff, depending on irrigation technology and management [3]. Irrigation systems are important in agriculture due to the need of dispensing water efficiently, even more because of the implications that climate change brings on with respect to the water availability. As explain in [4], ecological criteria are used for the definition of administration decisions in irrigation systems. These decisions depend on the amount of water required by a crop and availability in local sources as well [5].

Automatic irrigation controllers are the most rational approach to optimize irrigation, which aims to achieve the desired soil moisture level for each crop, improving costs and energy consumed in irrigation systems [6]. In automatic irrigation systems, quantity of water applied and the consequent crop response must be sensed in order to improve irrigation management, using historical data and control techniques [7].

The Smart Water Application Technology (SWAT) committee - Irrigation Association, defines *Smart controllers* as those technologies that "Estimate or measure depletion of available plant soil moisture in order to operate an irrigation system along with replenishment of water as needed while minimizing excess water use. A properly programmed smart controller requires initial site specific set-up and will make irrigation schedule adjustments, including run times and required cycles, throughout the irrigation season without human intervention" [8].

There are four important processes in automatic irrigation applications: data acquisition, interpretation, control and evaluation.

1.1   Data Acquisition

The temperature, soil moisture and solar radiation are environmental parameters which play an important role in the development process of plants [9]. The soil moisture, which is the amount of water contained in the soil pores, can be measured in terms of the volumetric water content (VWC), that means, the volume percentage of water present in a certain volume of soil [10]. Water contained in the soil intervenes as a reagent in photosynthetic and hydrolytic processes and serves as a solvent for salts, sugars and other solutes necessary for the cell growth of plants [11]. Hence, irrigation is an essential agricultural task to increase crop yield [12]. Current technologies facilitate the access and collection of relevant information about crops through the use of low cost sensors and simple wireless communication architectures. Several research projects have been developed that use sensors for irrigation management, especially soil, plant and atmosphere parameters technologies:

- Irrigation using soil water balance scheduling (evapotranspiration and rainfall estimation) [8]. In [13] a system was developed based on ZigBee communication, where, temperature, soil moisture, air humidity and the light intensity variables were measured, in order to implement better automatic irrigation control systems [14].
- Irrigation using soil moisture sensors (SMS), basically: soil water measurements (soil water tension (SWT) or soil water content (SWC) [15]. In [16], soil temperature and soil moisture data was collected at five different depths, with the aim of making future studies related to the physiology of the plants, and in [17] a reliable database was built from the information gathered by a sensor network.
- Scheduling irrigation using plant-based measurements (dendrometer, xylem cavitation, sap flow sensors, tissue water status or stomatal conductance) [18].
- Scheduling irrigation using spectral information [19].
- Scheduling irrigation using farmer knowledge [20].

1.2 Interpretation

Information and communication technologies (ICTs) in irrigation applications, require sensor integration using a computational system or software. This system has to be able to acquire data in real time, analyze it, and take decisions. In irrigation management this means that the software has to decide when to irrigate and how much water to apply. For this purpose the software could consider: irrigation water quality, cultivation system, plant stage, irrigation system, electrical conductivity (EC), drainage, nutrient requirements, solar radiation, vapor pressure deficit, and ambient temperature, among others [21]. An useful tool to obtain better performance of the irrigation activities is a decision support system (DSS), that is able to estimate if some process (irrigation) is required or not [6]. The implementation of artificial intelligence in decision support systems is used to interpret and analyze information acquired from sensors in the field, making irrigation application more scientific and accurate [22].

Last years have been created strategies to irrigate in different scales, using procedures that involve the development of prescription maps and automatic irrigation infrastructures as central pivots and drip irrigation systems. One of the main elements in automatic irrigation is the inference system, which allows to know procedures that will have to be deployed on the field. Support irrigation decision systems based on artificial intelligence advances, also called intelligent decision support systems, use some strategies to infer needs and procedures to water crops. For a machine to exhibit intelligence, it has to interpret and analyze the input and result data apart from simply following the instructions on that data. Some techniques and AI fields used in irrigation are [23]: Machine learning [24], knowledge engineering [25], fuzzy logic [26], artificial neural networks [27], case-based reasoning [28], expert systems [29], bayesian networks or probabilistic networks [30], artificial life [31], evolutionary computation [32] (genetic algorithms [33], evolutionary strategies [34] and evolutionary programming [35]), data mining or KDD (Knowledge Discovery in Databases) [36] and distributed artificial intelligence [37].

1.3 Control and Evaluation

An automatic controller can apply irrigation to refill water used by plants during several days, using crop potential evapotranspiration (ETc), changes in the soil water content or plant-based measurements. On the other hand, irrigation decisions can be taken based on heuristics, expert knowledge or a model of the system. Open-loop controllers presents some limitations that can be overcome by the use of feedback (closed-loop controllers), mathematical models and additional information provided by plant measurements [38]. The most deployed method of irrigation control is the closed-loop which splits into two categories: feed-forward and feedback control. In feedback control, the idea is to maintain soil moisture within a specific range by measuring crop's needs from soil moisture levels, using instruments such as tensiometers or dielectric probes. However, in the feed-forward control (known as ET control), controllers use the crop's reference evapotranspiration (ETo) to schedule irrigation, compensating then for ET water losses through the water balance technique. Climatic conditions have direct influence on ETo, which can be calculated by using Penman Monteith model as this has been officially adopted by the FAO [14].

Some control strategies that are able to switch on/off the irrigation pump and to open or close the valves to apply the irrigation doses to every sector of the crop are:

- **On-Off control.** Consisting on switching the controller output between maximum and minimum output according to an error signal. Examples of this kind of strategy are: soil water content [39]; canopy temperature [40]; soil matric potential, [41] and sap flow [42].
- **PID (Proportional-Integral-Derivative) control.** An example using soil water content is shown in [42].
- **Fuzzy logic control.** Fuzzy logic interprets real uncertainties and becomes ideal for nonlinear, time-varying and heuristic knowledge to control a system. Examples using soil water content, air temperature and light intensity in [43], using soil measurements (moisture and temperature) and solar radiation in [14], with multi-agent architecture in [44].

- **MPC (Model predictive control).** There are few agricultural applications (mainly for regulating weather conditions in controlled environments such as greenhouses) because it is difficult to obtain precise models appropriated for control purposes; however, it is a promising methodology for the design of irrigation controllers. An example using soil water content in [42].
- **Non linear control.** Nonlinear control theory focuses on systems that do not obey the superposition principle (linearity). These systems are often governed by non-linear differential equations. An example is in control methodology for irrigation distribution systems [45].
- **Artificial neural networks.** ANN are not really controllers, but a modeling framework which can be used in advanced model based controllers. Examples using soil water content in [46] and Back-Propagation Neural Network (BPNN) in [19], using Neural Network and Fuzzy Logic in [47].

1.4 Intelligent Agent

Artificial intelligence studies how to make computers and robots to behave or think in the same way as humans. Some branches of artificial intelligence research strategies which allow computers to reason according to mechanisms closest to those of humans. Others study methodologies which help machines and computers to carry out actions that a human being could develop using his own intelligence. Several AI techniques seek the development of models that allow rational thinking, while others use concepts of rational behavior, which means that computers or robots can perform actions considered as the best possible option to achieve the objectives for which they were programmed. This behavior is strongly related to the development of intelligent agents.

An intelligent agent is an autonomous entity, which acquires information from the environment using some kind of sensor, uses reasoning procedures and learning strategies to infer actions in the environment and carries out these actions using actuators. It is said that an agent is rational if the actions executed in the environment where it is found, are appropriate according to the response caused to its environment [48].

In this paper a fuzzy logic mechanism used to infer irrigation quantities and timing in crop systems is described. That mechanism is part of the intelligent agent rational behavior. This paper is organized by sections: methods and materials, design and configuration of the system, discussion of results, acknowledgments and conclusions.

2 Materials and Methods

This work was developed at the USOCHICAMOCHA irrigation district, located in Boyacá, Colombia. The district is divided into eleven zones, each one with its own pumping station. The water source is the Chicamocha River. The pilot farm is supplied by San Rafael pumping station, placed in the municipality of Nobsa. In this field are available: irrigation infrastructure, a weather station and water supply. The soil characteristics of the farm are: cultivation: kikuyo, soil: 14.10% clay, 42.80% silt and 43.10% sand, bulk density: 1.24, total porosity: 52.16, saturation point: 52.56%, field capacity: 36.60% and permanent wilting point 15.92%.

General scheme of the irrigation agent developed is showed in Fig. 1. Three hardware modules were developed as part of the intelligent agent: *Monitoring Station*, which allows the acquisition of soil moisture data; *Central Station*, considered the brain of the agent, where information is stored, analyzed and where inference system was implemented; *Irrigation station*, which is the system in charge of acting on the solenoid valves for irrigation tasks. Additionally, stored metrics can be monitored in a web server application and data is sent using a GPRS module.

The agent acquires information from sensors and according to its base of knowledge and its inference procedure based in fuzzy logic, establishes the irrigation assignments. Inference procedure uses the information coming from soil moisture sensors acquired in a monitoring station, as well as temperature and luminosity acquired from a weather station, to define how much time should last the irrigation task.

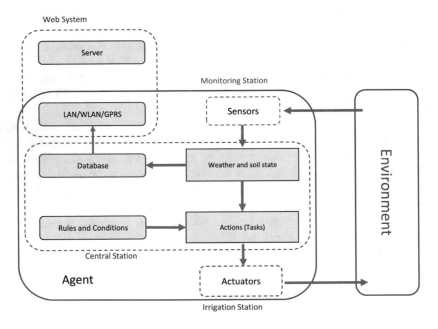

Fig. 1. Agent diagram. Source: Authors

2.1 Monitoring Station

Each monitoring station is composed of a soil temperature sensor (THERM200) and two soil moisture sensors (VH400), which measure the amount of water in the soil at two different depths. Each sensor has its respective signal conditioning stage, which is captured by an MSP430G2553 microcontroller that allows the information to be sent wirelessly by means of an XBee configured as *End Device AT*, Fig. 2.

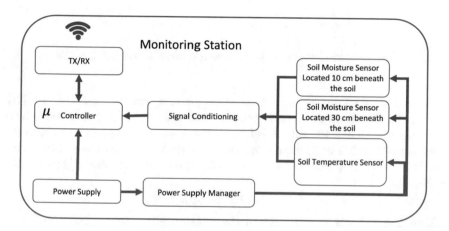

Fig. 2. Monitoring station diagram. Source: Authors

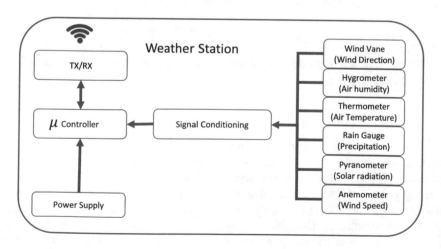

Fig. 3. Weather station diagram. Source: Authors

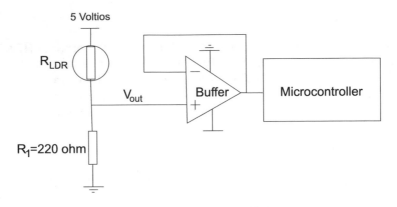

Fig. 4. Schematic diagram light sensor circuit. Source: Authors

A weather station was used to measure air temperature and solar radiation lighting (Fig. 3). A low-cost photometric sensor was used to estimate the behavior of day lighting due to the sun. A voltage divider and a voltage follower were used at the signal conditioning stage (Fig. 4). The output voltage of this circuit (V_{out}) reflects a value that can be associated with the solar radiation lighting follows the Eq. (1).

$$V_{out} = 5 * R_1/(R_{LDR} + R_1) \tag{1}$$

Equation (2) was used to obtain the relationship between *Light* and R_{LDR} [14]. Then, using Eqs. (1) and (2), Eq. (3) was obtained, that is useful to define light values.

$$log_{10}(R_{LDR}) = -0.5 * log_{10}(Light) + 4.23 \tag{2}$$

$$Light = 10^{8.46 - 2*log_{10}(\frac{215 - 220*Vout}{Vout})} \tag{3}$$

The weather station sends data autonomously and periodically (each 5 min) to the central station.

2.2   Irrigation Station

This module can manage two solenoid valves through the use of a ATMEGA328P micro-controller and XBee modules for wireless communication. *DC Latching Solenoids* (Interlock DC Solenoid) were used, which means that they can be activated by pulses. To change the valve state (open or closed), a voltage with reverse polarity regarding the previously applied pulse, must be deployed. Solenoid valves were handled by the L298N *driver* and a relay module. In Fig. 5, the configuration scheme of the designed irrigation station is presented. Figure 6 shows the software GUI page intended to expose scheduled irrigation tasks history.

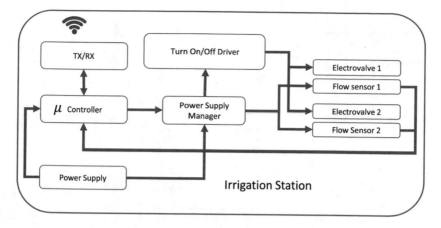

Fig. 5. Irrigation station diagram. Source: Authors

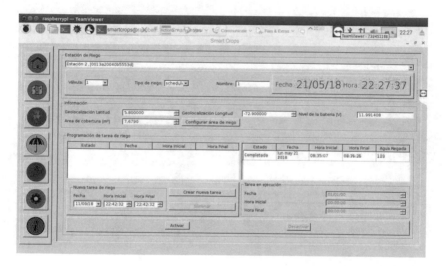

Fig. 6. Raspberry Software GUI, Irrigation Station Tasks History. Source: Authors

2.3 Central Station and Inference System

The central station is in charge of data storage and network management (Fig. 7). It consists of one XBee module configured as *Coordinator API*, a Raspberry Pi that acts as the agent information processing and analysis system, where wireless network settings and important parameters can be established by the user through a software application written in Python and Qt. This software

Weather Station
Monitoring Station 1
Monitoring Station 2
Monitoring Station n
Irrigation Station 1
Irrigation Station 2
Irrigation Station m
TX/RX
Central Station
Processing Hardware
System software (Controller and Manager)
Power Supply
Peripherals
GUI
Database
Network Interface
Web Server

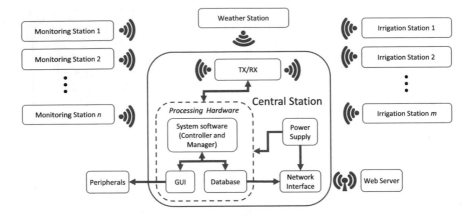

Fig. 7. Central station diagram. Source: Authors

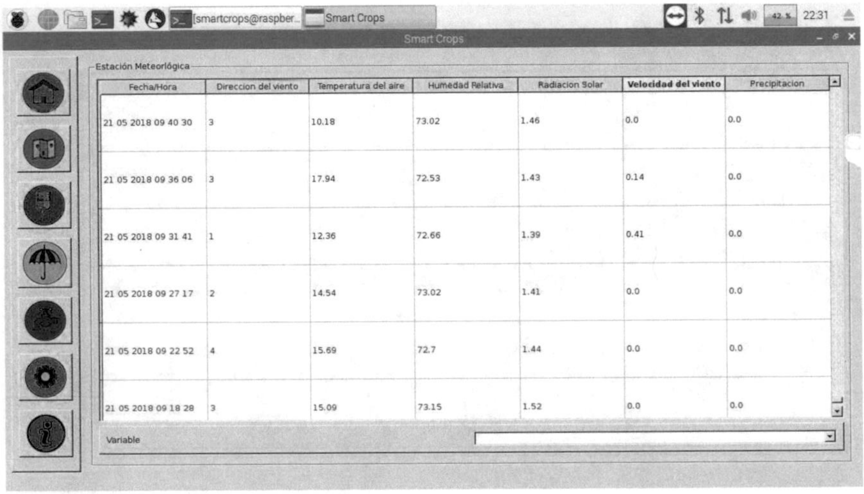

[smartcrops@raspber...
Smart Crops
22:31
Smart Crops
Estación Meteorlógica
Fecha/Hora
Direccion del viento
Temperatura del aire
Humedad Relativa
Radiacion Solar
Velocidad del viento
Precipitacion
21 05 2018 09 40 30 3 10.18 73.02 1.46 0.0 0.0
21 05 2018 09 36 06 3 17.94 72.53 1.43 0.14 0.0
21 05 2018 09 31 41 1 12.36 72.66 1.39 0.41 0.0
21 05 2018 09 27 17 2 14.54 73.02 1.41 0.0 0.0
21 05 2018 09 22 52 4 15.69 72.7 1.44 0.0 0.0
21 05 2018 09 18 28 3 15.09 73.15 1.52 0.0 0.0
Variable

Fig. 8. Raspberry Software GUI, weather station data page. Source: Authors

embed the fuzzy-logic-based inference system which was developed using the SciKit-Fuzzy library [49]. It is also in charge of requesting data periodically to the monitoring stations and sending asynchronous tasks to the irrigation stations. Figure 8 shows an screenshot for the weather station data page on the software GUI.

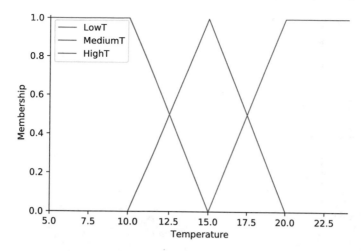

Fig. 9. Membership functions - Input: Temperature (°C). Source: Authors

Three inputs and one output were defined for the controller. Inputs are: environmental temperature, soil moisture and light intensity, the latter associated with solar radiation. For each input variable were used three membership functions and the output variable was defined using five membership functions related with the irrigation time.

Temperature values defined between 0 °C and 25 °C, which include the average of minimum annual temperature (9 °C) and the average of the maximum annual temperature (20 °C) in the region of study, were used in the universe of discourse for the temperature variable, Fig. 9. Values between 0 and 2500 lx, were used in the universe of discourse of brightness; these values were data obtained from the weather station installed in the field, Fig. 10. Finally, the third input, soil moisture, was defined with saturated, field capacity and permanent wilting points, between 15.92% and 52.56% VWC, whereby, the universe of discourse was set using values between 0% and 60% VWC, as a generalization of the measurement range, Fig. 11.

Temperature and light intensity membership functions were implemented using triangular and trapezoidal functions, defining three functions, each one with the center in the third part of the universe of discourse. On the other hand, soil moisture membership function is not symmetric respect to the universe of discourse, due to remarkable variable features. Table 1 specifies the rules used in the fuzzy logic algorithm (Fig. 12).

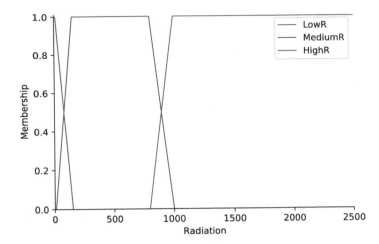

Fig. 10. Membership functions - Input: Brightness (Lux). Source: Authors

Table 1. Rules defined in inference system. Source: Adapted of [14]

Soil moisture (%VCW)	Temperature (°C)	Ilumination (Lux)		
		LowR	MediumR	HighR
LowM	LowT	VeryLong	VeryLong	VeryLong
	MediumT	Long	Long	Little
	HighT	VeryLong	VeryLittle	Nothing
MediumM	LowT	Little	Little	Little
	MediumT	Little	Little	VeryLittle
	HighT	Long	VeryLittle	Nothing
HighM	LowT	Nothing	Nothing	Nothing
	MediumT	Nothing	Nothing	Nothing
	HighT	Nothing	Nothing	Nothing

The inference system sub-process runs as follows: the central station ask the linked monitoring station for the current soil moisture value. The request interval time used was 5 min. When the captured value is below a threshold, established using the Eq. (4) [50],

$$FLC_{thr} = \frac{FC - PWP}{2} \tag{4}$$

where FC is the field capacity, PWP is the permanent wilting point and FLC_{thr} is the aforementioned threshold — the irrigation controller checks if there is no irrigation tasks currently running, and in that case the software builds an input

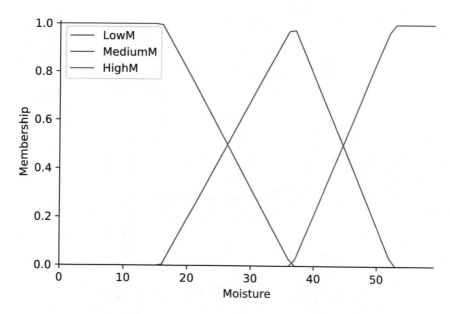

Fig. 11. Membership functions - Input: Moisture (% VWC). Source: Authors

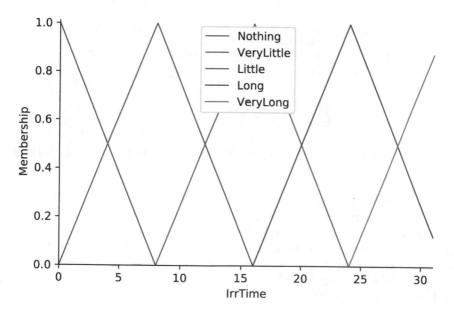

Fig. 12. Membership functions - Output: Irrigation Time (min). Source: Authors

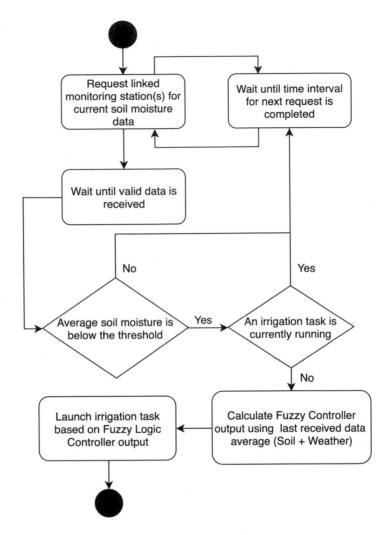

Fig. 13. Program subprocess diagram for irrigation using FLC. Source: Authors

array, taking the average of the last parameters values acquired in 30 min (Soil Moisture, Solar Irradiance, Air Temperature) and the *ControlSystemSimulation* class, defined using the previously described membership functions, calculates the inference system output and queues a time-controlled irrigation task with the duration corresponding to the FL controller evaluation. The forefront task will be launched asynchronously and processed by the irrigation station as soon as it can. This sub-process diagram is shown in Fig. 13.

3   Results and Discussion

Outcomes of the inference system are specified in Fig. 14, 15, 16, 17 and 18. In Fig. 14, a fuzzy-logic-based inference system output for a test input data is showed. System inputs were temperature: 15 °C, brightness: 500 lumens and soil moisture: 30% VWC. Using Mamdani strategy, the system obtain 18.73 min of irrigation. A simulation model of water balance was developed to assess system behavior under real conditions, using data previously collected from the weather station. In addition, in the simulation, a water depletion model was implemented to establish the water behavior in the soil and consumption in plants. Using an initial value of moisture, temperature and brightness of the weather station, the system defines the behavior of irrigation application in the field.

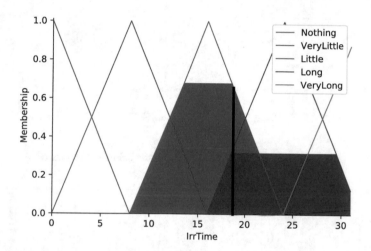

Fig. 14. Test fuzzy logic irrigation controller. Source: Authors

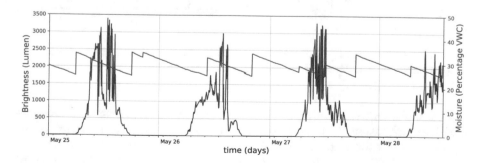

Fig. 15. Outcome - brightness and moisture. Source: Authors

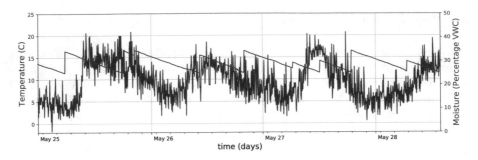

Fig. 16. Outcome - temperature and moisture. Source: Authors

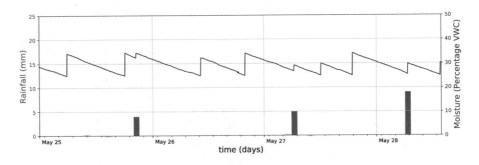

Fig. 17. Outcome - rain and moisture. Source: Authors

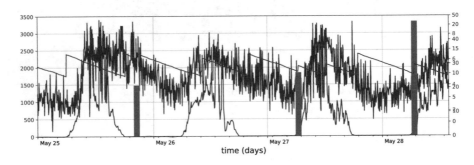

Fig. 18. Total outcome. Blue: Soil Moisture ($\%VWC$), Red: Brightness (Lux), Purple: Temperature (°C) and Green: Rainfall (mm). Source: Authors

Figure 15 shows the relationship between brightness and changes in soil moisture. The application of irrigation was developed when the solar radiation is low, in order to avoid the water evaporation, which is a behavior associated with the application of irrigation when the temperature is low, as shown in the Fig. 16. Figure 17 shows the behavior of the system with rain episodes. There are three rain events in the simulation period. For each event the system works according to the rules defined in the inference system. Figure 18 shows the relationship between all the variables used in the inference system, whose behavior was established according to the rules in Table 1. In that picture can be seen as well, that the main objective of the intelligent agent; which is in this case to keep the soil moisture close to the field capacity $(30\%VWC)$ and thereby, ensure the correct development of the plants; was achieved.

4   Conclusions

A wireless system for irrigation management was designed and implemented, taking as a reference the concept of rational agent. In this approach an inference system was implemented in the agent using fuzzy logic, which figures out when and how much water should be deployed in an agricultural crop, based on data gathered from soil moisture sensors and a weather station, and then, when required, the agent launches time-controlled irrigation task in order to keep plants under suitable conditions for their evolution.

Inference system explained in this document uses brightness, temperature and soil moisture sensors, which are useful to determine the interval of time necessary to maintain soil moisture at the point of field capacity, avoiding also the losses by evapotranspiration that they can be generated during daytime hours with high temperatures and solar radiation levels.

Only one soil moisture sensor was used to establish the irrigation application criteria. In future work, it is intended to use sensors at different depths and other types of sensors to determine the water needs of a crop using measurements in plant, soil and the environment.

The efficiency of water use is a defined parameter to relate the yield of the crop and the water applied. With the methodology developed in this work, runoff is avoided by excess irrigation of saturated soils in addition to allowing irrigation at adequate hours avoiding excess evapotranspiration, which allows to improve the efficiency of water use. The intelligent system proposed in this document makes the irrigation process more efficient by maintaining the necessary level of water in the soil so that the plants develop correctly and the excess of water applied is avoided.

Acknowledgment. We would like to thank the Research and Extension Department of Universidad Nacional de Colombia, Bogotá, for funding the institutional project entitled: Modeling Based on Agents for Precision Agriculture Applications in the National Call for Projects for research strengthening, Creation and Innovation of Universidad Nacional de Colombia 2016–2018. Pedro-F Cardenas expresses its gratitude to Colciencias for the doctorate study abroad scholarship 2007. Andres-F Jimenez expresses its gratitude to Crop, Soil and Environmental Sciences Department, Precision Agriculture, Auburn, Alabama, the Boyacá government for the doctorate scholarship - 2015 and also to the Universidad de los Llanos. We express our gratitude to the engineer Angel Rafael López Corredor.

References

1. Bendre, M., Thool, R., Thool, V.: Big data in precision agriculture: weather forecasting for future farming. In: 2015 1st International Conference on Proceedings of the Next Generation Computing Technologies (NGCT), pp. 744–750. IEEE (2015)
2. Mrinmayi, G., Bhagyashri, D., Atul, V.: A smart irrigation system for agriculture based on wireless sensors. Int. J. Innov. Res. Sci. Eng. Technol. **5**, 6893–6899 (2016)
3. Döll, P.: Impact of climate change and variability on irrigation requirements: a global perspective. Clim. Change **54**(3), 269–293 (2002)
4. Zhemukhov, R.S., Zhemukhova, M.M.: System of mathematical models to manage water and land resources at the regional level in case of anthropogenous climate changes taking into account economic indicators and ecological consequences. In: IEEE Conference on Quality Management, Transport and Information Security, Information Technologies (IT&MQ&IS), pp. 256–261. IEEE (2016)
5. Riediger, J., Breckling, B., Svoboda, N., Schröder, W.: Modelling regional variability of irrigation requirements due to climate change in Northern Germany. Sci. Total. Environ. **541**, 329–340 (2016)
6. Isern, D., Abelló, S., Moreno, A.: Development of a multi-agent system simulation platform for irrigation scheduling with case studies for garden irrigation. Comput. Electron. Agric. **87**, 1–13 (2012)
7. Smith, R.: Review of precision irrigation technologies and their applications. University of Southern Queensland, Technical report (2011)
8. Nautiyal, M., Grabow, G.L., Miller, G.L., Human, R.L.: Evaluation of two smart irrigation technologies in Cary, North Carolina. In: Proceedings of the Conference: 2010 Pittsburgh, Pennsylvania, 20 June–23 June, 2010, vol. 1. American Society of Agricultural and Biological Engineers (2010)
9. Lehnert, M.: Factors aecting soil temperature as limits of spatial interpretation and simulation of soil temperature. Acta Universitatis Palackianae Olomucensis–Geographica **45**(1), 5–21 (2014)
10. Datta, S., Taghvaeian, S., Stivers, J.: Understanding soil water content and thresholds for irrigation management. Oklahoma Cooperative Extension Service BAE-1537. Division of Agricultural Sciences and Natural Resources, Oklahoma State University (2017)
11. Mweso, E., de Bruin, S.: Evaluating the importance of soil moisture availability, as a land quality, on selected rainfed crops in Serowe area, Botswana. ITC (2003)
12. Mihailovic, B., Cvijanovic, D., Milojevic, I., Filipovic, M.: The role of irrigation in development of agriculture in srem district1. Ekonomika Poljoprivrede **61**(4), 989 (2014)
13. Chavan, C., Karande, P.: Wireless monitoring of soil moisture, temperature & humidity using zigbee in agriculture. Int. J. Eng. Trends **11**(10), 493–497 (2014)
14. Touati, F., Al-Hitmi, M., Benhmed, K., Tabish, R.: fuzzy logic based irrigation system enhanced with wireless data logging applied to the state of Qatar. Comput. Electron. Agric. **98**, 233–241 (2013)
15. Sui, R., Fisher, D.K., Barnes, E.M.: Soil moisture and plant canopy temperature sensing for irrigation application in cotton. J. Agric. Sci. **4**(12), 93 (2012)
16. Majone, B., Viani, F., Filippi, E., Bellin, A., Massa, A., Toller, G., Robol, F., Salucci, M.: Wireless sensor network deployment for monitoring soil moisture dynamics at the field scale. Procedia Environ. Sci. **19**, 426–435 (2013)
17. Terzis, A., Musaloiu-E, R., Cogan, J., Szlavecz, K., Szalay, A., Gray, J., Ozer, S., Liang, C.J., Gupchup, J., Burns, R.: Wireless sensor networks for soil science. Int. J. Sensor Netw. **7**(1–2), 53–70 (2010)

18. Grashey-Jansen, S.: Optimizing irrigation efficiency through the consideration of soil hydrological properties–examples and simulation approaches. Erdkunde, pp. 33–48 (2014)
19. Hendrawan, Y., Murase, H.: Neural-intelligent water drops algorithm to select relevant textural features for developing precision irrigation system using machine vision. Comput. Electron. Agric. **77**(2), 214–228 (2011)
20. Merot, A., Bergez, J.E.: Irrigate: a dynamic integrated model combining a knowledge-based model and mechanistic biophysical models for border irrigation management. Environ. Model. Softw. **25**(4), 421–432 (2010)
21. Rodriguez-Ortega, W., Martinez, V., Rivero, R., Camara-Zapata, J., Mestre, T., Garcia-Sanchez, F.: Use of a smart irrigation system to study the eects of irrigation management on the agronomic and physiological responses of tomato plants grown under different temperatures regimes. Agric. Water Manag. **183**, 158–168 (2017)
22. Chen, Z., Liu, G.: Application of artificial intelligence technology in water resources planning of river basin. In: 2010 International Conference of Information Science and Management Engineering (ISME), vol. 1, pp. 322–325. IEEE (2010)
23. Bustos, J., Ricardo, J.: Inteligencia artificial en el sector agropecuario. Seminario de Investigación I. Universidad Nacional de Colombia, Colombia (2005)
24. Kaur, K.: Machine learning: applications in Indian agriculture. Int. J. Adv. Res. Comput. Commun. Eng. **5**(4), 342–344 (2016). https://doi.org/10.17148/IJARCCE.2016.5487
25. Rafea, A., Hassen, H., Hazman, M.: Automatic knowledge acquisition tool for irrigation and fertilization expert systems. Expert Syst. Appl. **24**(1), 49–57 (2003)
26. Zhang, Q., Wu, C.H., Tilt, K.: Application of fuzzy logic in an irrigation control system. In: Proceedings of the IEEE International Conference on Industrial Technology, ICIT 1996, pp. 593–597. IEEE (1996)
27. Zanetti, S., Sousa, E., Oliveira, V., Almeida, F., Bernardo, S.: Estimating evapotranspiration using artificial neural network and minimum climatological data. J. Irrig. Drain. Eng. **133**(2), 83–89 (2007)
28. Li, X., Yeh, A.: Multitemporal SAR images for monitoring cultivation systems using case-based reasoning. Remote. Sens. Environ. **90**(4), 524–534 (2004)
29. Fedra, K.: Models, GIS, and expert systems: integrated water resources models. In: Proceedings of the International Conference on Applications of Geographic Information Systems in Hydrology and Water Resources Management, Vienna, vol. 211, pp. 297–308. IAHS Press (1994)
30. Castelletti, A., Soncini-Sessa, R.: Bayesian networks and participatory modelling in water resource management. Environ. Model. Softw. **22**(8), 1075–1088 (2007)
31. Dessalegne, T., Nicklow, J.W.: Artificial life algorithm for management of multi-reservoir river systems. Water Resour. Manag. **26**(5), 1125–1141 (2012)
32. Ranjithan, S.R.: Role of evolutionary computation in environmental and water resources systems analysis (2005)
33. Wardlaw, R., Bhaktikul, K.: Application of genetic algorithms for irrigation water scheduling. Irrig. Drain. **53**(4), 397–414 (2004)
34. Reddy, M.J., Kumar, D.N.: Evolving strategies for crop planning and operation of irrigation reservoir system using multi-objective differential evolution. Irrig. Sci. **26**(2), 177–190 (2008)
35. Pant, M., Thangaraj, R., Rani, D., Abraham, A., Srivastava, D.K.: Estimation of optimal crop plan using nature inspired metaheuristics. World J. Model. Simul. **6**(2), 97–109 (2010)

36. Khan, M.A., Islam, M.Z., Hafeez, M.: Evaluating the performance of several data mining methods for predicting irrigation water requirement. In: Proceedings of the Tenth Australasian Data Mining Conference, vol. 134, pp. 199–207. Australian Computer Society, Inc. (2012)
37. Le Bars, M., Attonaty, J.M., Pinson, S.: An agent-based simulation for water sharing between different users. In: Proceedings of the First International Joint Conference on Autonomous Agents and Multiagent Systems: Part 1, pp. 211–212. ACM (2002)
38. Romero, R., Muriel, J., García, I., de la Peña, D.M.: Research on automatic irrigation control: state of the art and recent results. Agric. Water Manag. **114**, 59–66 (2012)
39. Boutraa, T., Akhkha, A., Alshuaibi, A., Atta, R.: Evaluation of the effectiveness of an automated irrigation system using wheat crops. Agric. Biol. J. N. Am. **2**(1), 80–88 (2011)
40. O'Shaughnessy, S., Evett, S.R.: Canopy temperature based system effectively schedules and controls center pivot irrigation of cotton. Agric. Water Manag. **97**(9), 1310–1316 (2010)
41. Romero, R., Muriel, J., Garcia, I.: Automatic irrigation system in almonds and walnuts trees based on sap flow measurements. In: VII International Workshop on Sap Flow, vol. 846, pp. 135–142 (2008)
42. Vicente, R.R.: Hydraulic modelling and control of the soil-plant-atmosphere continuum in woody crops. (doctoral dissertation). (2011). http://hdl.handle.net/10261/96715
43. Xiang, X.: Design of fuzzy drip irrigation control system based on zigbee wireless sensor network. In: International Conference on Computer and Computing Technologies in Agriculture, pp. 495–501. Springer (2010)
44. Salazar, R., Rangel, J.C., Pinzón, C., Rodríguez, A.: Irrigation system through intelligent agents implemented with arduino technology. Adv. Distrib. Comput. Artif. Intell. J. (ADCAIJ) **2**(3), 29–36 (2013)
45. Benayache, Z., Besançon, G., Georges, D.: A new nonlinear control methodology for irrigation canals based on a delayed input model. In: IFAC Proceedings Volumes, vol. 41, no. 2, pp. 2544–2549 (2008)
46. Capraro, F., Patino, D., Tosetti, S., Schugurensky, C.: Neural network-based irrigation control for precision agriculture. In: IEEE International Conference on Networking, Sensing and Control, ICNSC 2008, pp. 357–362. IEEE (2008)
47. Tsang, S., Jim, C.: Applying artificial intelligence modeling to optimize green roof irrigation. Energy Build. **127**, 360–369 (2016)
48. Russell, S.J., Norvig, P.: Artificial Intelligence: A Modern Approach. Pearson Education Limited, Malaysia (2016)
49. Warner, J., Sexauer, J., scikit-fuzzy, twmeggs, Unnikrishnan, A., Castelō, G., Batista, F., Badger, T.G., Mishra, H.: Jdwarner/scikit-fuzzy: Scikit-fuzzy 0.3.1 (2017). https://doi.org/10.5281/zenodo.1002946
50. Osroosh, Y., Peters, R.T., Campbell, C.S., Zhang, Q.: Comparison of irrigation automation algorithms for drip-irrigated apple trees. Comput. Electron. Agric. **128**, 87–99 (2016)

Decision Support System for Precision Irrigation Using Interactive Maps and Multi-agent Concepts

Giovanny Hernández[1], Andres Fernando Jimenez[1,2,3(✉)], Brenda V. Ortiz[3], Alfonso P. Lamadrid[2], and Pedro Fabian Cardenas[1]

[1] Unrobot Research Group, Universidad Nacional de Colombia, Bogotá, Colombia
{gihernandezga,pfcardenash}@unal.edu.co

[2] Dynamic Systems Research Group, Universidad de los Llanos,
Villavicencio, Meta, Colombia
{ajimenez,aportacio}@unillanos.edu.co

[3] Crop, Soil and Environmental Sciences, Precision Agriculture,
Auburn University, Auburn, AL, USA
bvo0001@auburn.edu

Abstract. Agriculture is a fundamental pillar in the economy, sustainability and food security. To achive a correct and efficient management of crops, there are strategies, which use the integration of different technologies currently found in engineering, especially electronics and software, to develop methodologies based on measurement, decision support and action on crops. Precision agriculture (PA) is one of those techniques, and it is a strategy used to determine when, where and how to apply the inputs on crops. Geographic Information Systems (GIS) are tools that can be used in PA for the information management of agricultural fields, where, interactive maps allow the characterization of crops, soils and to define prescriptions through Decision Support Systems (DSS). In this work, an useful tool to obtain soil moisture maps of agricultural fields and a water distribution model in USOCHICAMOCHA irrigation district located in Boyacá- Colombia, using a GIS and multi-agent concepts are shown. A negotiation strategy based on the Contract Net protocol is activated when there is a scarcity of water resources, defining priorities for the irrigation distribution in a more efficient manner.

Keywords: Interactive · Geographic information system
Precision agriculture · Multi-agent

1 Introduction

Agriculture is one of the fundamental axes in the development of a region and is a vital factor in environmental sustainability, poverty reduction and food security. For the efficient management of agricultural crops, different strategies and technologies have been developed, among which precision agriculture is highlighted. Through the use of geographic information systems, wireless networks

© Springer Nature Switzerland AG 2019
J. C. Corrales et al. (Eds.): AACC 2018, AISC 893, pp. 21–41, 2019.
https://doi.org/10.1007/978-3-030-04446-3_2

and specialized software technologies, researchers and producers seek to improve the efficiency in the application of inputs, in such a way that they are used in the right place, at the right time and in the right amount [1].

Precision agriculture is a term that describes the use of different techniques like geospatial information, GPS systems and sensing, in order to identify changes in field conditions for attending crop needs, using different strategies. This concept has now taken strength given its importance in several society areas and is precisely the result of the need to optimize and manage natural resources [2,3].

Likewise, AP is expected to improve economic competitiveness in the agricultural sector, showing better results in product quality with a lower production price than the current one. This is possible through process technification, the constant measurement, observation and action on the crop. This also opens the way of increasing potential production and product diversification for a better price and quality, due to the fact that cropland losses can be predicted [4].

Because of these reasons, precision agriculture and engineering, converge to give a reasonable use and management of natural resources, including land treatment and water consumption. Thus, engineering seeks not only to provide a solution to current generations problems, but also seeks, to conserve and generate friendly solutions with the environment. Therefore, this development must look for a long-term solution instead of a short-term solution [5].

Research in favor of the environment, the suitable and effective use of the natural resources seems to be the current worldwide scenario. In consequence, it has generated great interest on the variations of the biogeophysics processes through its temporary and spatial variability [6]. Remote sensing and low cost technologies of soil moisture sensors provide an effective way to know crop conditions at regional level. Thus, a better manipulation and application of the supplies and treatments could be possible, implying that every crop receives the precise quantity of water [7].

At this point it is important to highlight the United Nations Food and Agriculture Organization (FAO) statistics, which shows that 50% of water extraction is used to irrigate agricultural crops in Colombia [8]. This means a high waste of water distribution in the country. The situation could be worst if we take into account that some places have not access to water nowadays [9].

Due to the fact, that Colombia has a great diversity, agriculture is a sector that possesses an enormous weight in the country economy. According to National Administrative Department of Statistic (DANE), 7.6% of the soil was used for agriculture purposes in Colombia at 2015. Because of that, research and investment, national and international, has increased in the last years [10].

Nowadays, there are some solutions to automatic crop irrigation relying on a rule-based approach, which often uses soil moisture, supported by an unlimited water supply. Moreover, these solutions take advantage of predictive applications which tend to be problematic in places and years where there is not information or water resource availability [11]. Additionally, there are some solutions for monitoring specific phenomena information [12,13] but none of them integrates agent-based technologies (artificial intelligence) with the use of sensors and visualization (monitoring) in real time, specifically for colombian crop irrigation systems.

In irrigation, the environmental impact produced by water consumption and possible waste of it, could be reduced using new technologies. Artificial intelligence is an option to define water application in crops, and intelligent agents are proposed in this research to define irrigation criteria in agricultural systems. An intelligent agent is an entity that seeks to achieve some objectives in an environment in which interact, using some available resources, with the ability of take decisions with the collaboration with another entities, that is known as a multi-agent system [14–16]. A multi-agent system is related with networks of intelligent entities, that exist within an environment, that can communicate information using the network, in which nodes are artificial intelligence systems. The agents behavior, together with others entities, produces an intelligent group result. Agents are partially autonomous, with local visualization of the environment, wherein no exist a control agent designed. Those systems can manifest autoorganization, autodirection, another control paradigms and complex behaviors related.

The Study approach of multi-agent systems is in analysis and solution of sophisticated problems of artificial intelligence and control architectures. Can be divided in different kind of technologies: homogeneous robot multi-agent systems [17] and heterogeneous software multi-agent systems [18]. Each one has communication technologies between agents, decision mechanisms, data acquisition from the environment, other collaborative agents, decision support systems and learning methodologies.

A model of collaborative multi-agent system for water management in agricultural crops is shown in this paper. The principal aim of the system is to contribute to the development of colombian technologies useful in farming management. This paper is divided in sections. In section two, materials and methods are disclosed, in section three, results and discussion are appreciated, in section four, acknowledges are established and finally in section five, conclusions are defined.

2   Materials and Methods

USOCHICAMOCHA irrigation district, located in Boyacá, Colombia, was used as example to show the model behavior explained in this paper. The district is divided into eleven zones, each one with its own pumping station. The water source is the Chicamocha River. Pilot fields of this study belong to Monquirá pumping station, that is located near to the municipalities of Sogamoso and Nobsa. In this zone are available: irrigation infrastructure, a weather station and water supply. Crops types are: corn, pastures and onion. Soil characteristics of the region are defined in values between: 1.41%–58.49% clay, 27.56%–75.71 % silt and 5.72%–38.16% sand, field capacity: 36.44%–61.29% and permanent wilting point 16.68%–57.24%. Those values define the variability in the study region.

With the implementation of a plugin in Quantum GIS, is possible the use of sensors and remote sensing images to define water requirements. With this information, a coordinator agent takes the role of manager in an auction, field agents send information about water requirements, and the coordinator, according with its knowledge (map information), could define the veracity of the agent request and can take decisions about the fields that could apply irrigation. As result, a model that integrates spatio-temporal information of croplands using spatial data in vectorial format and multi-agent negotiation could be developed. Figure 1 shows the principal components of this model.

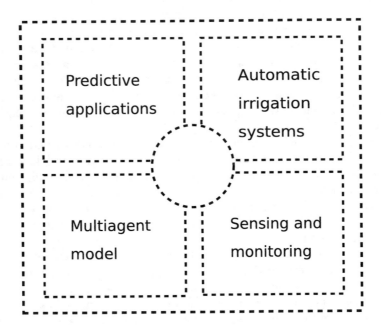

Fig. 1. Project components diagram. Source: Authors

The decision support system takes into advantage the synergy of four main fields with the purpose of maximize the use of water resource, these components involve a multi-perspective view to analyze and take decisions for the fair distribution of the water, following, it will be explained the principal approach of each component:

2.1   Predictive Applications

The analysis and modeling of crops behaviors take throw a way of correlate some variables that strongly affect the consumption of water, in the way that the complexity of the model is increased and the interaction established between variables predict with certain accurate the state of the crop, the model will be stronger. With the parametric model of crop behavior, it can be predicted the state of the crop for some seasons and in this way take actions about it.

The decision support system integrate this method into the model for predict soil evapotranspiration that take a direct influence in soil moisture and crop health. According to the temperature and the relative humidity in a region, the evapotranspiration will vary, this causes crop yield losses, water waste and a high cost in material losses.

2.2 Automatic Irrigation Systems

The progress in farming technologies has taken the inevitable route of automating its processes. Due to the needs of taking care of the agricultural ecosystem, different techniques are being developed, especially the control of some variables in the environment. The regular climate change and the extreme conditions in the land make this tasks have an elevated level to carrying out, for this reason the inclusion of new technologies have taken a high importance in crop irrigation.

An automated irrigation system can be as simple or complex as desired, but the principle is the same, it is to have a controlled environment for the soil; this is accomplished by adding some temporal actions in which the crop is irrigated according to environmental variables such as soil moisture.

For this purpose the decision support system uses a monitoring station in each crop at certain distance for taking information about the current soil conditions and determine the action on it. Actuators on the field are mainly composed by a solenoid valve system to control the opening and closure of water supply to the crop.

2.3 Multi-agent Model

Sometimes, due the limited restriction of water resources, it is not possible to irrigate the crops with complete water requirement every day. The search of the common great and the optimization of this resource have to take into account multiple agents (crops) that can interact between them with the purpose of obtain an optimum distribution of water resource; this is the reason because a multi-agent model is useful in this task.

A multi-agent *Contract Net* cooperation protocol is used as water negotiation mechanism, to assign irrigation scedules to fields, based in the offer and demand of a service [15]. In this protocol, exists two kind of agents: The **Manager**, role assumed by district irrigation administration, that is in charge to administrate the water, and coordinate water distribution to other agents called **Contractors**, that are agents that use sensors on the fields to define soil water status. In Fig. 2 the manager agent sends messages to every contractor informing about the availability of the resource; every contractor that needs water answer with an offer (Fig. 3), then, the manager determine the priority of irrigation application (Fig. 4) and answer with a denial or acceptance message to each contractor agent (Fig. 5).

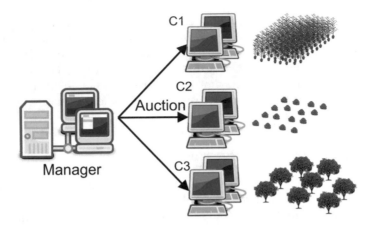

Fig. 2. Water resource auction - Manager. Source: Authors.

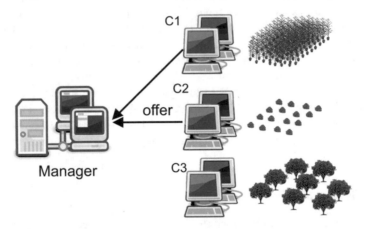

Fig. 3. Contractors' offer. Source: Authors.

Following this procedure, irrigation could be applied on fields that received an acceptance message. With this model, a multi-agent system is used to define irrigation schedule with multiple fields that use a same water supply. Contractor agents are located and acquire field information, that allows them to define irrigation activities independently on each location, using sensors. When water resource is not enough, negotiation procedure is activated and the manager agent is in charge of coordinate this activity. For this reason manager agent needs to know information about water needs in fields to take decisions; for such purpose, spatio - temporal information of the region could be used to define decision criteria to select fields with water priority needs. With this model is possible to define water requirements in crops using a spatial vectorial base, where each crop is monitored by an agent as a poligon feature.

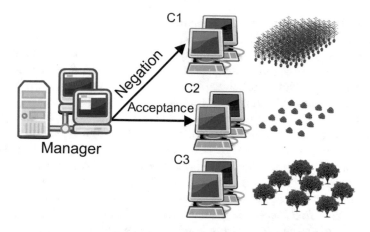

Fig. 4. Manager response to the offers. Source: Authors.

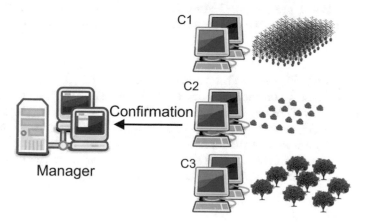

Fig. 5. Confirmation. Source: Authors.

Water availability conditions and field needs can be analyzed according to the following conditions, that allow to define rules used in *Manager* and *Contractor* nodes. To simplify, water available is represented as a reservoir (excavation made in field to store irrigation water) using a tank (four stacked blocks), useful to define four states: *ResLevel1*, alarm in availability of the resource (only one block); *ResLevel2*, resource availability level A, minimum limit available for irrigation (two available blocks); *ResLevel3*, resource availability B (three blocks) and finally *ResLevel4*: maximum capability in the reservoir (four blocks). Water available in the reservoir can't be less that the threshold defined.

In irrigation is important take into account crop characteristics, soil moisture and environmental conditions. In this study crop water needs are defined according to the measurements in soils. Four fundamental states are associated with soil water content: permanent wilting point - PWP (limit to obtain permanent damage of plants), tolerance level (maximum limit to avoid permanent wilting point), field capacity (soil moisture adequated) and saturation (maximum soil water retention). In the first approximation, is assumed that for change to each one of these states is applied the same quantity of water, ergo, to change from permanent wilting point to tolerance level is applied the same quantity of water, as if the change is from field capacity to saturation level.

Rules definition in *Contractor* agent, due to water needs in each field are established by means of a tank with four states: *FieldLevel1*, soil moisture is in PWP. *FieldLevel2*, corresponds to the tolerance level and must necessarily be watered to avoid reaching the PWP (two blocks available), *FieldLevel3*, represents field capacity, that is, if the crop is within *FieldLevel2* and *FieldLevel3*, there is the most appropriate (ideal) water availability condition (up to three available blocks); *FieldLevel4*, four full blocks represent saturation. For rules definition in this work, last level is not used, because the drainage management is not taken into account and also, it is a state in which no amount of water needs to be applied. Below are the possible situations in the contract net protocol for a *Manager* and two *Contractors*, associated to fields. Rules are defined according to water availability.

Situation 1. Two Fields in the Tolerance Limit. As a first situation, there is complete availability of water resources, *ResLevel4* (Fig. 6), then the Manager starts the auction. There are two agents in the field that are in the minimum limit *FieldLevel2*, so they respond, each one requesting irrigation to the Manager. The manager knows field states and establish that both need water according to the measurements in the field. In this way, each agent have the possibility of apply irrigation, to keep the best condition (field capacity).

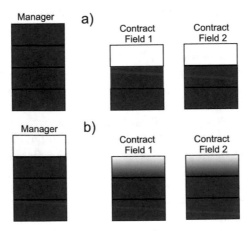

Fig. 6. Situation 1. Two Fields in the tolerance limit. Source: Authors.

Situation 2. A single Field with Critical Level. In this case, there is water availability, *ResLevel3* (Fig. 7). One of the fields has a condition associated with PWP, that is, a critical state with high priority level (*FieldLevel1*), then after the resource availability auction, the agent in the field answers the request, specifying its level of water content. The Manager only receives this offer and verifies the need, applying the amount of water to take the batch to the state of the best condition (field capacity).

Situation 3. Possibility of Risk of PWP with Scarce Availability of Water Resources. The amount of water in the reservoir is sufficient for a certain irrigation application, without affecting its capacity for subsequent applications (*ResLevel3*). The Manager starts the auction and in this case two agents respond stating that the monitored crops are at risk of PWP (*FieldLevel1*). The Manager analyzes the information and decides to apply irrigation to both crops without reaching its minimum status, because the water source can't be put at risk (Fig. 8).

Situation 4. No Water Resources Available. In this situation, there is no water necessary to carry out an auction because the reservoir is within the allowed limit (*ResLevel2*), so that the crops in the fields may reach a state of PMP (*FieldLevel1*), there is no way to apply irrigation, (Fig. 9).

In Table 1, states corresponding to each of the possible actions in the system are appreciated. The initial states s change to the state s', after having made the action a, $(s' = result(s, a))$. These are the rules used in the Manager to be able to make water distribution in the most limited conditions of available amount of water in the reservoir, in the case of two cultivated fields. L represents the minimum limit of the reservoir and p is the maximum value of water content that each lot must have.

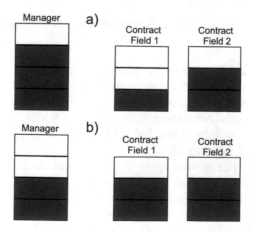

Fig. 7. Situation 2. A single Field with critical level. Source: Authors

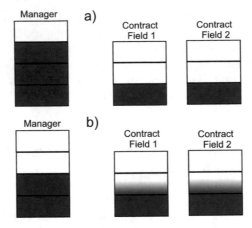

Fig. 8. Situation 3. Possibility of risk of PWP with scarce availability of water resources. Source: Authors.

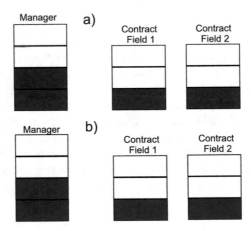

Fig. 9. Situation 4. No water resources available. Source: Authors

A multi-agent model that uses the contract net protocol explained previosly, allows to communicate water requirements from each individual agent (crop) to ponder and maximize the distribution of water resources among the crops. Data transmission allows to the manager to know water needs from each agent, and the relationship of multiple agents helps to know the environmental behavior, converting it into a help network.

The decision support system apply this model for crop irrigation, that combined with the predictive model and the use of sensor and automated irrigation systems performs a high accurated perspective to the optimization of the water resource for irrigate crops. The complexity of the model and the autonomy of the system increase and strengthen the probability of an efficient irrigation.

Table 1. Initial and final states according to design. Source: Authors.

Initial states s		Final states s'	
Reservoir (R)	Contractors (ag)	Reservoir (R')	Contractors (ag')
$R \leq L$	ag_i	$R' = R$	$ag'_i = ag_i \forall i$
$L + 1 < R < L + 2$	$ag_1 = ag_2 = p$	$R' = R$	$ag'_i = ag_i \forall i$
$R = L + 1$	$ag_1 = ag_2 = p - 2 \text{ o } ag_1 = ag_2 = p - 1$	$R' = R - 1$	$ag'_i = ag_i + 0.5$
$R = L + 1$	$ag_1 = p - 2, ag_2 = p - 1$	$R' = R - 1$	$ag'_1 = ag_1 + 1,$ $ag'_2 = ag_2$
$R = L + 1$	$ag_2 = p - 2, ag_1 = p - 1$	$R' = R - 1$	$ag'_2 = ag_2 + 1,$ $ag'_1 = ag_1$
$R = L + 1$	$ag_1 = p - 2, ag_2 = p$	$R' = R - 1$	$ag'_1 = ag_1 + 1,$ $ag'_2 = ag_2$
$R = L + 1$	$ag_2 = p - 1, ag_1 = p$	$R' = R - 1$	$ag'_2 = ag_2 + 1,$ $ag'_1 = ag_1$
$R = L + 2$	$ag_1 = ag_2 = p - 2 \text{ o } ag_1 = ag_2 = p - 1$	$R' = R - 2$	$ag'_i = ag_i + 1 \forall i$
$R = L + 2$	$ag_1 = p - 2, ag_2 = p - 1$	$R' = R - 2$	$ag'_1 = ag_1 + 1.5,$ $ag'_2 = ag_2 + 0.5$
$R = L + 2$	$ag_2 = p - 2, ag_1 = p - 1$	$R' = R - 2$	$ag'_2 = ag_2 + 1.5,$ $ag'_1 = ag_1 + 0.5$
$R = L + 2$	$ag_1 = p - 2, ag_2 = p$	$R' = R - 2$	$ag'_1 = ag_1 + 2,$ $ag'_2 = ag_2$
$R = L + 2$	$ag_2 = p - 2, ag_1 = p$	$R' = R - 2$	$ag'_2 = ag_2 + 2,$ $ag'_1 = ag_1$
$R = L + 2$	$ag_1 = p - 1, ag_2 = p$	$R' = R - 1$	$ag'_1 = ag_1 + 1,$ $ag'_2 = ag_2$
$R = L + 2$	$ag_2 = p - 1, ag_1 = p$	$R' = R - 1$	$ag'_2 = ag_2 + 1,$ $ag'_1 = ag_1$

2.4   Sensoring and Monitoring

The constant evaluation of field conditions makes possible to obtain a complex system to analyze crops, and the quality of measurement variables will have an impact in the model prediction. The precision and accuracy of these instruments affect the behavior of the complete system. On the other hand, the rate of sensing and data transmission are important in the performance of the system.

For this reason is important to have a monitoring station that constantly acquires data from the enviroment and acts based on them. In a similar way, it is important to have an easy way for reading this data using an graphic interface, due the amount of data per second that is obtained in each station. The monitoring is an important activity to maintain and evaluate field conditions in an easy understanding way for human inspectors.

2.5   Model Integration

Integrating these four perspectives is possible to create new rules and methods for irrigation, taking into account the advantages of each technique to strength a general model. Each perspective separately has its advantages and disadvantages, some of them do not take into account some important variables for their model to work correctly. Further, some of them prioritize the crop itself but not their neighbors and some of them assume there is an unlimited availability of water resource.

Control and monitoring of a greater number of variables improves prediction and action on soil conditions. These analysis and actions also require a great amount of resources and processing capabilities, where there is a greater challenge when defining in some places, equipments and instruments that the system could use according to the availability of budget for its implementation.

Using a global perspective of field conditions, water resources will be distributed in an equitative way, where each crop has the essential conditions to survive and the water will no be wasted or used incorrectly.

An adecuated tool for monitoring make easier to the farmer to understand the state and behavior of his crops, due the importante of the correct reading of the state of the crops, the monitoring application have to be intuitive and quick to understand.

3   Results and Discussion

In this paper, the integration of spatio-temporal information of a crop region in Boyacá (Colombia) is carried out, through the creation of the spatial vectorial base, where information of sensors in field (agents) is visualized, and used to establish the procedure to support decisions of the Manager agent in the model. For this purpose, a plugin in Quantum GIS was developed. The project is composed of three fundamental pillars as will be explained below.

3.1   Vector Layers in QGIS

In order to obtain a graphical user interface according to the information that has been obtained and an intuitive visualization, it becomes necessary the generation of geographical irrigation crop maps, in which monitoring stations are implemented. For this purpose, Quantum GIS (QGIS) software was used, which is useful for the management and treatment of maps and information in big quantities. In QGIS was generated a vector layer that contains spatial information of fields that allows to know each cropland condition. This Layer is modified according to the information obtained from a web server, where remote sensing images or wireless sensor networks are the principal data sources, for the coverage area. Every crop has an identification number (ID) which is used as descriptor object number in QGIS, to have concordance between the information and the visualization. A plugin was implemented in QGIS software for monitoring these crops, it allows the user to obtain field variables, for example temperature or humidity, at time intervals defined in the plugin.

3.2   Data Acquisition Modules

It is necessary to know several variables that intervene in crop development, since the actions that are carried out in these, depend on the plant, the soil and the environment. In irrigation, sensors are an excellent technique to know the variability of soil moisture in a field, and the information acquired could be used in irrigation control systems to achieve moisture levels in the soil, near to the field capacity value. To manage information in field, an intelligent agent design was performed to acquire information, process the data and activate electrovalves using prescriptions, Fig. 10. Intelligent agent is simulated to obtain data useful in the QGIS plugin to know measurements of soil moisture in fields belonging to Monquirá region of USOCHICAMOCHA irrigation system.

Field Agent or contractor, uses soil moisture and soil temperature to define water requirements. For water application in fields, actuators are used, in this case solenoid valves, that allow open and close the water flow, according with crop needs. Communication between several field agents modules and

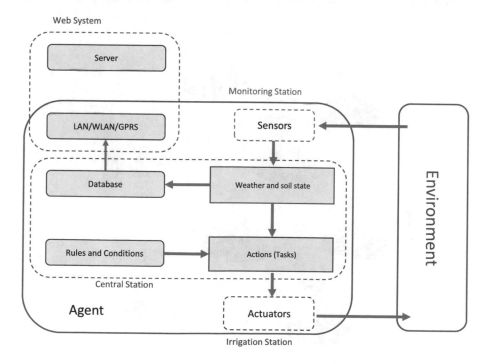

Fig. 10. Agent diagram module. Source: Authors

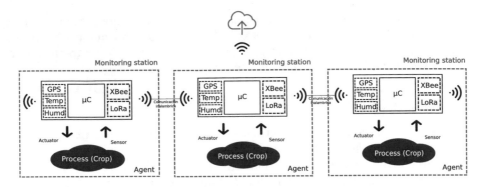

Fig. 11. Data acquisition modules scheme. Source: Authors.

central station uses XBee Pro S2C communication boards, which are wireless communication devices of low power consumption. Each agent acquire information for one crop or field and send this values to a Web-Server. Figure 11, shows a representation of some data acquisition modules.

A more complex networks make possible a multi track for the spreading of information, this have a double purpose, due the large extension of the lanes no all the communications systems have the range to cover those limits, taking into account that everymodule works with solar energy the modules need to be of

Fig. 12. Data propagation in the network. Source: Authors.

low consume, and in other hand for a better evaluation of the field conditions it is good to cover not so large areas, so the communications modules have a good relation of range and consumption this also help to propagate the information among agents in a close area and create a chain effect until the information arrives to the central station (Fig. 12). By other hand, if every agent is aware of its neighbours, in case some of them fault their neigbours can alarm to the central of this fault. The constant communication between this agents is an important factor in the model.

3.3   Integration System for Visualization in QGIS

Finally, for the development of the *Manager* Decision Support System, connection between data acquisition modules and QGIs plugin is required. This procedure is developed using the cloud, given the versatility that this solution offers for acquiring data from anywhere in the world through multiple devices and technologies (Fig. 13).

Looking for a solid platform, this is created based on a microservices architecture. For that reason, a cloud server receives information from any device and deliver information to any device; this is made by a Representational State Transfer architecture (REST). Publisher devices make a call to the server giving the information in a JavaScript Object Notation (JSON) formated text, requesting for the information updated from a given crop state at any time. On the other

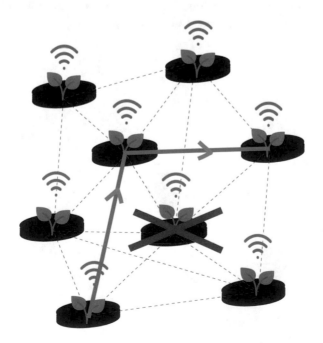

Fig. 13. Data propagation with a node in fault. Source: Authors.

hand, subscribers make a GET request to the server, this responds with a JSON formated text with the information of the crops. This make possible that any device can publish and subscribe to the web server information like a web script, a mobile app, a desktop software or even an embedded system.

The server host a PHP REST API than can be used by multiple devices and has three fundamental methods:

– **GET:** Return the information from the server. It posses an elevated security (always is run in the server side) and regardless of how many times it is repeated with same parameters, results are the same. Does not modify the database.
– **PUT:** Create or modify a register.
– **DELETE:** Request that a record has been deleted.

The client side or the subscriber is the QGIS application, that gets the JSON text response and update every crop information state. In monquirá region there are multiple crops, then, interactive maps are visualized in an intuitive way using QGIS Layer of the region, with a graduated color scale corresponding to field water requirements. *FieldLevel1*, is represented by red color (PWP), *FieldLevel2*, uses yellow color (Tolerance Level) and *FieldLevel3* is green (Field Capacity) (Fig. 14).

The simplicity in accessing this information becomes an indispensable factor, this is the reason why a QGIS plugin was developed. The strategy allows the user to select the layer (Map) to be linked with the server information (every crop have an identifier number ID that must correspond with the ID of every data

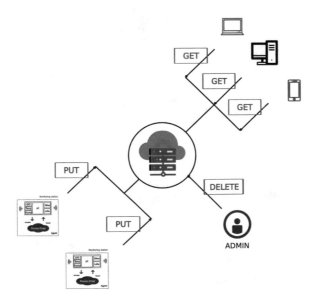

Fig. 14. Decision support system integrated systems scheme. Source: Authors.

Fig. 15. Screenview of the Plugin developed in QGIS. Source: Authors.

acquisition module in the field, also a GPS verification is made for a concordance between the QGIS Maps and data acquisition modules in the irrigation district. With this strategy, *Manager* agent has enough information to decide how take decisions about water distribution. In Fig. 15, the plugin user interface (UI) is presented.

When the plugin is working, update automatically field information about the moisture state and repaint the map according to the variable to be visualized. To illustrate the system behavior, moisture levels in Monquirá pump station (Boyacá, Colombia) were simulated, Fig. 16 shows crops states in this area, according to moisture levels defined in the multi-agent negotiation model.

Versatility and modularity of this software make possible to expand the system to multiple platforms and languages. The use of an universal and light weight formated text for data interchange increase the system performance. The appropriate elaboration of the database (normalizing the database tables) make possible a high efficiency in data acquisition and search. A multiservices architecture make possible to scale to a great rate, with an extraordinary modularity and minimizing the dependency among modules.

In this work, permanent monitoring in multiple platforms, historic information and decision support system implementation was developed to improve water use in irrigation. This system creates an historical database of information about spatio-temporal variations in fields for predictive applications. This aspect combined with a rule-based approach and a multi-agent model (justified in AI techniques) create a strong baseline for an automatic irrigation system.

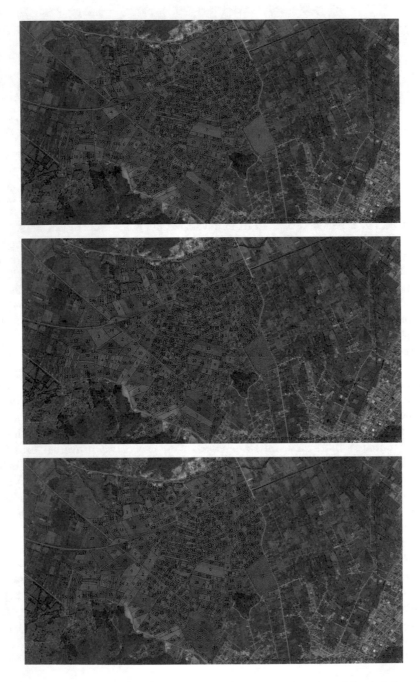

Fig. 16. Map examples of irrigation needs. Priority definition - Monquirá region - USOCHICAMOCHA. Red color (PWP), *FieldLevel2*, uses yellow color (Tolerance Level) and *FieldLevel3* is green (Field Capacity). Source: Authors.

4   Conclusions

In this research, a decision support system model was developed, in order to provide a tool for water distribution in irrigation districts, giving water prescriptions according to values obtained in fields and approximations to real crop needs. The system integrate multi-agent concepts to manage irrigation in a distributive way, when there is not enough water to supply homogeneously the resource to multiple users. In low water availability a multi-agent negotiation protocol was proposed to achive water distribution using needs of crops according to an auction procedure known as Contract Net protocol. The objective of the developed model is to avoid the excess or lack of water in the soils, trying to keep the soil moisture near to field capacity level.

There are two principal structures studied, the first one, related with a field irrigation management using the conception of an intelligent agent, which allows the acquisition of soil temperature and moisture, and the activation - deactivation of solenoid valves to apply irrigation. The second structure, is based on multi-agent modeling, which uses each crop or field as an intelligent agent. All agents seek to apply irrigation correctly in an agricultural region of USOCHI-CAMOCHA district irrigation, using a global perspective of field conditions. where the use of Manager agent and Contractors agents is possible to distribute the water in an informed manner.

The combination of multiple techniques improves the performance of the model, system complexity increases and factors that influence the process could be more controlled. An intuitive visualization improves the temporal response for taking decisions by farmers about their crops. The correct color used and map size in the user interface developed, make easier to the farmer, as Contract agent to understand the behavior and knows water needs in his crops. But the main benefit of managing the information in the vector layer is that, the associated table could be used by the Manager agent to define if a crop or field needs water and the amount approximated to administer the negotiation protocol.

An appropriate distribution of the water is relevant due the importance of this valuable resource, even more in all the industrial fields where in many scenarios is wasted or used incorrectly, given that this resource is not only use for irrigation the distribution needs to have a global perspective, a multi-agent model takes importance in this area, where each agent is aware of its own needs and their neighbours, the total amount of water is divided in such a way that the most quantity individues have the resources to survive. This implies a correct and efficient use of water, reducing in this way the possible lost of crops due an insuficient amount of water. This entails, an increment of earnings for farmers and a better care of the enviroment.

A multi agent model for optimize the distribution of water is a concept that can be extended to other fields, the water as other resources is limited and the importance for the care of the enviroment has to be principal concern, an efficient distribution of each resource could not only help the enviroment but also have a great economic repercusion. The model can be applied for many context where the measurement and control of the consuming of each resource is based on a logic of the common benefit in pro of optimize attending to the needs of this new era of changes.

Acknowledgement. We would like to thank the Research and Extension Department of the National University of Colombia, Bogota, for the financing of the institutional project entitled: Modeling Based on Agents for Precision Agriculture Applications, the National Call for Projects for the Strengthening of Research, Creation and Innovation of the National University of Colombia 2016–2018. Cárdenas PF expresses its gratitude to Colciencias for the doctorate study abroad scholarship 2007. Jiménez AF expresses its gratitude to the Boyacá government for the doctorate scholarship of the call for the formation of high-level human capital for the department of Boyacá - 2015 and also to the Universidad de los Llanos. We express our gratitude to the engineer Angel Rafael López Corredor.

References

1. González, C., Sepúlveda, J., Barroso, R., Fernández, F., Pérez, F., Lorenzo, J.: Idesia **29**(1), 59 (2011)
2. Zhang, C., Kovacs, J.M.: Precis. Agric. **13**(6), 693 (2012). https://doi.org/10.1007/s11119-012-9274-5
3. Wachowiak, M.P., Walters, D.F., Kovacs, J.M., Wachowiak-Smolíková, R., James, A.L.: Comput. Electron. Agric. **143**, 149 (2017). https://doi.org/10.1016/J.COMPAG.2017.09.035. https://www-sciencedirect-com.ezproxy.unal.edu.co/science/article/pii/S0168169916309401
4. Katalin, T.G., Rahoveanu, T., Magdalena, M., István, T.: Procedia Econ. Financ. **8**, 729 (2014). https://doi.org/10.1016/S2212-5671(14)00151-8. https://www-sciencedirect-com.ezproxy.unal.edu.co/science/article/pii/S2212567114001518
5. McBratney, A., Whelan, B., Ancev, T., Bouma, J.: Precis. Agric. **6**(1), 7 (2005). https://doi.org/10.1007/s11119-005-0681-8
6. Torres, A., Gómez, A., Jiménez, A.: Sistemas y Telemática **13**(33), 27 (2015). https://doi.org/10.18046/syt.v13i33.2079. https://www.icesi.edu.co/revistas/index.php/sistemas_telematica/article/view/2079
7. Jimenez, A., Jimenez, F., Fagua, E.: Revista Colombiana de Tecnologias de Avanzada (RCTA) **1**(21) (2013). https://doi.org/10.24054/16927257.V21.N21.2013.291. http://revistas.unipamplona.edu.co/ojs_viceinves/index.php/RCTA/article/view/291
8. Margat, J., Frenken, K., Faurès, J.M.: Intersecretariat working group on environment statistics (IWG-Env). International work session on water statistics, Vienna (2005)
9. UNICEF, La infancia, el agua y el saneamiento básico en los planes de desarrollo departamentales y municipales. In: Ronderos, M.T. (ed.) UNICEF 2015, 1st edn. Chap. 3, pp. 31–55 (2015). https://www.unicef.org/colombia/pdf/Agua3.pdf
10. Departamento Administrativo Nacional de Estadística (DANE), Encuesta Nacional Agropecuaria ENA-2016. Technical report, Bogotá (2016). https://www.dane.gov.co/files/investigaciones/agropecuario/enda/ena/2016/boletin_ena_2016.pdf
11. Winter, J.M., Young, C.A., Mehta, V.K., Ruane, A.C., Azarderakhsh, M., Davitt, A., McDonald, K., Haden, V.R., Rosenzweig, C.: Environ. Model. Softw. **96**, 335 (2017). https://doi.org/10.1016/J.ENVSOFT.2017.06.048. https://www.sciencedirect.com/science/article/pii/S1364815217300014

12. Monika, Srinivasan, D., Reindl, T.: 2017 IEEE Conference on Energy Internet and Energy System Integration (EI2), pp. 1–6 (2017). https://doi.org/10.1109/EI2.2017.8245766
13. Kainthura, P., Singh, V., Gupta, S.: 2015 1st International Conference on Next Generation Computing Technologies (NGCT), pp. 584–587. IEEE (2015). https://doi.org/10.1109/NGCT.2015.7375188. http://ieeexplore.ieee.org/document/7375188/
14. Ferber, J.: Multi-agent System: An Introduction to Distributed Artificial Intelligence. Addison Wesley Longman, Harlow (1999)
15. Gonzalez, E.: Robótica cooperativa Experiencias de sitemas multiagentes (SMA), 1st edn. Editorial Pontificia Universidad Javeriana (2012)
16. Solórzano, G.A.P.: Plataforma de desarrollo para multiagentes robóticos. Ph.d. thesis, Uniandes (2007)
17. Candea, C., Hu, H., Iocchi, L., Nardi, D., Piaggio, M.: Robot. Auton. Syst. **36**(2), 67 (2001)
18. Li, H., Karray, F., Basir, O.A., Song, I.: JACIII **10**(2), 161 (2006)

IoT Architecture Based on Wireless Sensor Network Applied to Agricultural Monitoring: A Case of Study of Cacao Crops in Ecuador

Juan Carlos Guillermo[1]([✉]), Andrea García-Cedeño[1]([✉]),
David Rivas-Lalaleo[2]([✉]), Mónica Huerta[1]([✉]), and Roger Clotet[3]([✉])

[1] GITEL, Telecommunications and Telematics Research Group,
Universidad Politécnica Salesiana, Cuenca, Ecuador
{jguillermo,agarciac,mhuerta}@ups.edu.ec
[2] Universidad de las Fuerzas Armadas ESPE, Sangolquí, Ecuador
drrivas@espe.edu.ec
[3] Networks and Applied Telematics Group, Universidad Simón Bolívar,
Caracas, Venezuela
clotet@usb.ve

Abstract. This article proposes the design of a low-cost IoT architecture based on Wireless Sensor Network technology for agricultural monitoring which is deployable in different types of crops, in this particular case cacao crops have been chosen. The architecture will allow small and medium agricultural producers, through a multi-platform application (web, mobile, etc.), to monitor and store information of various climatic and soil factors that influence the optimal growth and production of cacao. The data collected by the sensor network will be visualized through interactive maps, tables and statistical graphs. Furthermore, the system will notify or emit alerts to the farmer through emails or the application's own graphical interface about a specific event that has occurred, in order to support optimal and opportune decision-making. The proposal of this architecture corresponds to a inclusive tool of the agro-technological field for family cacao farmers through the collection, storage, management and visualization of agricultural variables data. The information acquired is fundamental for the sustainable management of the crop through the correct administration of resources, facilitating the quality control of the product and the generation of preventive plans for protection against pests and diseases.

Keywords: Wireless Sensor Network · Monitoring · IoT · Agriculture

1 Introduction

Agriculture has been one of the main activities within the primary sector of production, and simultaneously in social development and environmental

J. C. Corrales et al. (Eds.): AACC 2018, AISC 893, pp. 42–57, 2019.
https://doi.org/10.1007/978-3-030-04447-3_3

preservation. It is the key agent in food security and biodiversity conservation through the sustainable management of environmental resources [3,7]. At an economic standpoint, this activity plays an essential role in the development of nations because of its practice at a family level in the rural sector, consolidating its progress. In Ecuador, it is the main income for rural areas [25], which comprise 37% of the Ecuador's population [31]. According to the latest World Bank statistics, agriculture in Ecuador accounted for 10.25% of GDP (Gross Domestic Product) in 2016 [18], and in 2017 it constituted 26.93% of total employment in the country [19].

Among the main Ecuador's agricultural productions for exportation, are bananas, shrimp and cacao, with Ecuador being the country with the highest production and exportation of fine aroma cacao, also called Cacao Arriba, whose quality is internationally appreciated. Similarly, CCN-51 cacao considered as conventional, is a cloned cacao created in Ecuador, widely used for being of high productivity and quality, as well as tolerant to diseases. Its cultivation stands out due it is intended for the production of processed chocolate products and which corresponds to 25% of cacao exports in general [5,6,26]. Data generated by the Continuous Agricultural Surface and Production Survey (ESPAC) indicate that in 2017, a total of 205,955 tonnes of cacao were produced and 203,368 tonnes were sold, representing exports of $588.4 million USD according to the Central Bank of Ecuador (BCE) [6,21]. Ecuador is positioned worldwide among the five countries that grow this fruit [4] with more than 400 years of experience [26].

With the adaptation of an agricultural society to the modern world, for small and medium producers who practice family farming, sustainable development at the environmental and productive level becomes a challenge [25]. This problem also affects cacao farmers, who together generate national production and whose socioeconomic reality significantly affects Ecuador's agriculture. Among its main problems are low productivity, marketing obstacles and land distribution. In the productivity sphere, diseases such as Monilia and Witch's broom are the main factors of damage since they can generate losses of more than 50% of the crop [11,27], and because of their low capacity to combat them due to lack of capitalization, it is not possible to invest in the care and improvement of plantations at the technological level despite the existence of low cost systems [27]. It has been demonstrated that multiple agricultural technology solutions, especially those focused on information gathering and processing through the use of Wireless Sensor Network (WSN), are effective in increasing production and optimal use of resources compared to traditional farming methods [29,30].

In search of solutions to the problems presented in relation to family farmers, an Internet of Things (IoT) architecture based on WSN is proposed, applied to agricultural monitoring of cacao crops in Ecuador, which constitutes the initial phase of the project named MOCCA, Cacao Monitoring System, focused on reducing low productivity through the use of technological tools. Considering the socio-economic conditions of the farmers, it has been established that this proposal prioritizes low cost since open source software and hardware have been used.

This system does not intervene in an invasive way with natural development of the crop and its biodiversity, its importance is manifested in the optimization of the cultivation process, eliminating sequential maintenance practices in which the use of resources is planned according to time periods instead of sowing needs. With the respective monitoring of agricultural variables, the requirements of the plantations will be determined with precision, reducing the consumption of water, fertilizer, agrochemicals and the labor involved. The information obtained will be used to generate preventive plans for protection against pests and diseases.

This document is distributed as follows: in Sect. 2, the most relevant related work for the development of this article will be summarized. In Sect. 3, the methodology concerning the proposed architecture will be detailed. Section 4 is focused on the result. Finally, in Sect. 5, a discussion will be held.

2   Related Work

In the academic-research field, there is a variety of work focused on the application of technology in agriculture, which uses WSN as a tool and the interconnection of these platforms to the Internet to generate different projects under the concept of IoT. Within this classification, articles related to this thematic have been considered either on the basis of their purpose, architecture, or hardware used.

Giuseppe Aiello, in his article "A decision support system based on multisensor data fusion for sustainable greenhouse management", states that it is difficult for small farmers to accomplish green economic policies in Europe due to their socio-economical and cultural condition, therefore, he proposes a system based on a WSN designed from Arduino nodes and Zigbee modules, for the sustainable management of a greenhouse for horticultural production. The information acquired corresponds to environmental and plant parameters [2]. Another projects based on WSN with similar hardware corresponds to Culman, Ceballos and Erazo [10, 12, 13], which focuse in monitoring an oil palm plantantion, a habanero pepper crop and a rose greenhouse.

Ferrández-Pastor [14], develops a Wireless Sensor/Actuator Network platform implemented in a hydroponic greenhouse for irrigation control, the author evaluates soil and environmental variables made visible through a Web application and share the information obtained in Comma Separated Variable (CSV) files. Under the same objective focused on the correct administration of water, Flores-Medina [16], analyzes the soil moisture of a caladium greenhouse, making use of hardware and software from the company Libelium and visualizing the information through a web application. Similarly, Kodali [24], makes use of a WSN for programmed irrigation in response to the water stress produced during the dry season in coffee crops in India.

Emphasizing food safety, specifically the quality of agricultural products, different projects are born which aim to guarantee the good condition of food, such as the articles by Flores [15] and Karimi [23], which, based on WSN, perform measurements of environmental and soil variables through the use of Arduino boards and Zigbee communication devices, they are applied in open fields in a corn plantation and vineyards respectively.

A work using LoRaWAN (Long Range Wide Area Network) protocol, as is the case of this proposal, corresponds to Cambra [8], which consists in a WSN connected to an irrigation controller in conjunction with aerial image processing. It provides information about weather, ground conditions, and water flow and pressure, its data can be displayed through a mobile application.

There is documentation of agricultural technologies implemented in cacao plantations, however, most of them are based on the principle of spectrometry, image analysis and related techniques. In the case of a WSN application, there is the article "Cloud Computing for the Internet of Things. Case Study on Precision Farming" by Hernández [20], which corresponds to a case of study focused on information management through Cloud Computing to be applied to the irrigation control of a cacao field, but it does not consist on an experimental investigation. Another example where an IoT system is implemented in cacao production corresponds to Ipanaqué [22], in which, by means of sensors implemented in a device, they monitor temperature, oxygen and carbon dioxide variables for the analysis of the fermentation process of cacao beans.

Extensive research can be found in the area of WSN applied in agriculture, but very little specialized scientific evidence can be obtained in the area of cacao. This paper aims to expand this line of research by proposing an IoT architecture of an agricultural monitoring system, designed for small and medium-scale cacao producers, in other terms, a low-cost implementation with an user-friendly interface through a multiplatform application and therefore accessible from all means.

3 Development Methodology

Once the problems faced by small and medium-scale cacao farmers had been identified, dialogues were established with agronomists from the Ministry of Agriculture and Animal Husbandry of Ecuador (MAG) and family farmers, and a process was carried out to gather information on an initial crop. The cacao used in this study is of CCN51 type, also known as conventional, this classification stands out for its yield due to the fact of starting production after sowing in fewer years compared to Cacao Arriba, it has larger seeds, high butter content and is more resistant to fungal diseases [9]. An area of 1 hectare will be analyzed, located in Luz de América compound, 135 m above sea level, in the parish Molleturo, belonging to Azuay province (Ecuador), with the purpose of determining the climatic variables and physical and chemical parameters whose evaluation is necessary for the correct development of the fruit in question. The area in this case of study counts with internet access, which allows sending information to the servers located in Universidad Politécnica Salesiana in Cuenca (Ecuador).

According to the Ecuador's Agricultural Quality Assurance Agency (AGRO-CALIDAD) [1], the optimum temperature ranges for cacao cultivation are between 24–26 °C, the relative humidity tolerable maximum should not be greater than 60%, the soil pH should remain between values of 6–6.5 (neutral), a luminosity of 2000 h of light per year, rainfall and irrigation between 1500 and 2000 mm of water distributed throughout all months of the year.

The development methodology of the MOCCA system for cacao monitoring is structured under a client server type IoT architecture, which is divided into three layers: Application Layer, Service Layer and Sensor Layer as shown in the Fig. 1.

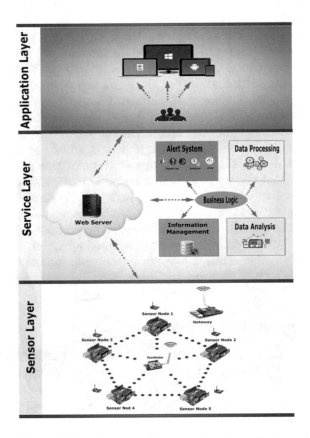

Fig. 1. IoT architecture for agricultural monitoring.

Through this architecture and the interaction of the three layers, small and medium agricultural producers will be able to: monitor and collect information on climate and soil variables and then store it, process it using Data Mining tools, techniques and algorithms and finally manage it through the Internet. The interface of the MOCCA system will allow the farmer to visualize the information collected from the WSN through tables, statistical graphs and interactive maps. In the same way, it will notify or generate alerts about a given event via emails, pop-ups, etc. The purpose of IoT architecture is to collect data, analyze it and make optimal and timely decisions by the system itself and the farmer. Each of the layers with their respective modules is described and detailed below.

3.1   Sensor Layer

This layer is in charge of acquiring and sending the data of the different climatic and soil variables involved in the growth and production of the cacao crop, through a WSN made up of five sensor nodes that operate under a mesh topology, a coordinator node and a gateway. The nodes in mention allow to monitor the cultivation of cacao, these sensors are characterized by being low cost due to these devices do not provide high accuracy or wide operating range, however their features are enough for this specific use. See Table 1.

Table 1. Hardware components of sensor nodes.

Component	Model	Technical data		
		Measure variable	Operating range	Resolution
Sensor	Hygrometer with protected probe	Soil moisture	0−100%	No spec
Sensor	DHT22	Environment humidity and temperature	0−100%RH −40 80°C	0.1%RH 0.1 °C
Sensor	All-sun Analogue pH meter for gardening	pH	1−10	No spec
Sensor	Homemade prototype	Electric Conductivity	No spec	No spec
Sensor	TSL2561	Luminosity	0.1−40,000 lux	0.1 lux
Single-board micro-controller	Arduino UNO SMD R3	Power consumption: 46 mA, 15 mA (sleep)		
Wireless communication module	XBee S2C	Communication protocol: Zigbee, Power usage: 45 mA Range: 120 m Line of Sight (LoS)		

Each sensor node processes and transmits its collected data to the coordinator node, using an XBee module based on the Zigbee wireless communication standard (IEEE 802.15.4). Sensor node hardware features are detailed in Table 2.

Subsequently, the coordinator node receives a data frame with data from the sensors through the LoRa module and sends it to the gateway, see its characteristics in Table 3. LoRa is a registered trademark of Semtech and consists of one

Table 2. Hardware components of coordinator node.

Component	Model	Technical data
Single-board microcontroller	Arduino UNO SMD R3	Power consumption: 46 mA, 15 mA (sleep)
Wireless communication module	XBee S2C	Communication protocol: Zigbee Power usage: 45 mA Range: 120 m (LoS)
Wireless communication module	Dragino LoRa GPS Shield for Arduino	Communication protocol: LoRaWAN (868MHz) Power usage: 10.3 mA Range: >10 Km (LoS)

of several ISM band options, with the main feature of offering long range and low power consumption. This wireless technology works under the LoRaWAN standard, corresponding to the protocol that incorporates LoRa as a physical layer [33]. The gateway through a LoRa module, receives the entire data frame and sends it via TCP (Transmission Control Protocol) to a WAMP Server on Internet, in which the Web application, the MySQL database and the Web Service are hosted.

Table 3. Hardware components of gateway.

Component	Model	Technical data
Single-board computer	Raspberry Pi Modelo B	Power consumption: 46 mA, 15 mA (sleep)
Wireless communication module	Dragino LoRa GPS HAT for Raspberry Pi	Communication protocol: LoRaWAN (868 MHz) Power usage: 10.3 mA Range: >10 Km (LoS)

3.2 Service Layer

It provides the necessary services for the farmer to manage, process, visualize the WSN data and generate alerts immediately. All these processes that involve the interaction between the Sensor layer and the Application layer are carried out through four modules: information management, alert management and data analysis/processing. All these services are hosted in the cloud to be able to access them remotely from any geographical location. Each of the modules is described and detailed below.

Information Management Module: By using this module, the administrator can: configure the WSN, record, modify and delete nodes and sensors,

consult and visualize the information collected by the WSN in tables, statistical graphs and interactive maps. In addition can download daily, monthly and
annual reports of historical data. However, the farmer can mainly see the current
data of the monitored variables of one or all the WSN nodes and also consult
the history.

The interaction with the network and services layer is achieved using an
intermediate layer of business logic programmed in PHP, JavaScript or Angular,
being the PHP Laravel Framework the tool used in the development of the
MOCCA system. WSN data will be stored in the MySQL Database (DB). The
PHP programming language was used in the web server to program the business
logic, basically because it has an open source license and several frameworks
that are compatible with many programming languages. As it is a collaborative
language, it offers support for the different inconveniences that may arise at
software and hardware level, since the 83% of all web sites use php as server-side
programming language [32].

This module seeks to provide the necessary information at the right time
to all actors and institutions involved in cacao cultivation. It also provides the
security of stored data by defining user profiles or roles.

Alert Management Module: It allows to control the maximum and minimum
levels of the climatic and soil variables of the cacao crop. It is very important
to manage these parameters because the control of their variations can sway in
opmitization of resources as well in prevent the proliferation of diseases such as
the monilla, black cob, witch's broom and others, which affect the production of
cacao. The maximum and minimum limits of each variable, previously specified,
are pre-established in the system with the possibility of editing them at any
time.

The following module that is described below, is a proposal that has been
taken into consideration for its implementation in the second stage, since the
complete data set obtained in the first monitoring phase is not available.

Data Analysis and Processing Module: It allows to preprocess the raw
data contained in the records of the five sensor nodes of a data source stored
in JSON format, in order to find and correct errors, eliminate duplicate and
inconsistent records, verify that the data does not contain incomplete records
(lack of attribute values), verify and solve noise problems (lost values), among
others. To obtain quality information, raw data acquired from a given data source
must go through the processes described in Fig. 2.

All these processes are achieved through data analysis tools such as: Mahout,
MLlib, FlinkML and H2O. Of these four tools, in this project we are evaluating the MLib Framework because it contains much more complete mass data
preprocessing algorithms that support tasks in the knowledge extraction process
such as classification, optimization, regression, grouping, and preprocessing. Also
because it stands out as the Framework that offers a greater variety of preprocessing methods in relation to the tools mentioned above [28].

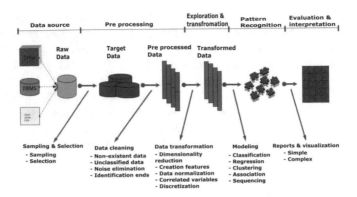

Fig. 2. Data mining process [28].

Once the data was processed, the system internally performs the distributed processing necessary to obtain reports and relevant information for the farmer using data mining processes through Frameworks such as Apache Spark, Flink and Hadoop. Of these three, the Apache Spark Framework has been chosen for evaluation because it has become one of the most powerful and popular tools in the Big Data ecosystem and also because it has proven to be more efficient than Hadoop and Flink in many cases of use. Thanks to its memory intensive operations, this platform is able to load data into memory and consult it quickly, making it a perfect tool for iterative processes where the same data is reused several times for processing algorithms on networks [17].

The analysis and processing module is mainly intended to improve the efficiency of the data mining process, enhance the relevant data selection, obtain data converted into quality information and reduce the size of the original data. It is necessary to indicate that this module is not yet implemented in the application, mainly because the different data mining techniques are being evaluated and also due to the fact that the application is designed and focused in first place on the monitoring of climatological and soil variables.

3.3   Application Layer

A multiplatform application has been designed that allows the farmer to interact with the IoT services that compose the system to remotely perform different actions. Within this variety of options highlights the configuration and control of WSN. Equally important the app offers the visualization of the information of the sensor network, a list of the sensors per node, historical data query (by sensor type, by a range of dates, by location of sensors through interactive maps), among others.

For the development of the application the IONIC Framework is used, a system that allows the creation of Android, IOS, Windows Phone and web browser applications, through programming languages such as HTML, CSS and Angu-larJS.

4   Results

An user friendly and intuitive multiplatform application based on IoT has been developed to provide technological support to small and medium agricultural producers with the purpose of collaborating mainly in the increase of cacao production, regulation of excess pesticides and fertilizers (optimization of resources), time and effort reduction, saving of costs in the harvest and above all supporting the making of optimal and opportune decisions about the crop. The MOCCA system allows access only to users registered and validated by the administrator. The home screen can be seen in Fig. 3.

Fig. 3. User authentication.

There are two types of roles assigned in the MOCCA system: the administrator and the owner of the crop. The administrator can mainly perform the following activities: add, modify and delete users, accept or reject user requests and add or remove options in the menu. In turn, he can modify content in the application, create, delete and edit nodes and sensors among others. On the other hand the owner of the crop will be able to check the current data and the history of each of the variables monitored by time intervals (daily, monthly and yearly). It is relevant to indicate that the administrator, unlike the owner of the crop, has an exclusive interface, as can be seen in the Fig. 4.

With the implementation of the five components of the architecture belonging to the service layer, it is possible to: monitor the variables, generate alert reports and visualize the historical data. Below they describe and detail each one of them.

As indicated in Figs. 5 and 6, the MOCCA system allows to monitor the climatological and soil variables, through pop-ups and statistical graphs that are capturing the WSN sensor nodes.

Thus it is also possible to generate alarm reports pertinent to events that occur in the nodes of the WSN. Keep controlled the maximum and minimum levels of temperature, humidity, etc (show Fig. 7).

Finally, as show in the Figs. 8 and 9, the system allows to visualize in statistical tables and graphs the historical data stored in the Mysql database by a

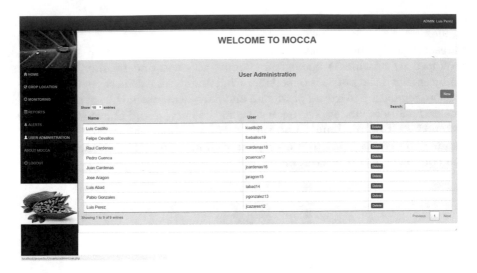

Fig. 4. Users administration

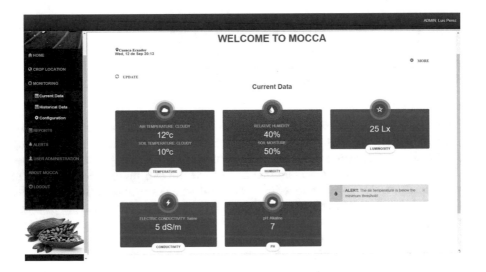

Fig. 5. Current data visualization through pop-ups.

range of determined dates (daily, monthly or annual), data of all the sensors or one in specific (temperature, humidity, etc) depending on the node in which it is installed.

With the information presented by the application the farmer, will be able to analyze and make decisions for the improvement of their crop. For example,

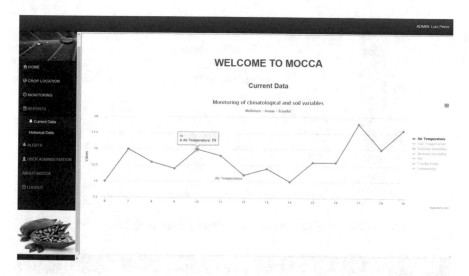

Fig. 6. Current data visualization through statistical graphs.

Fig. 7. Alert report.

identify the best time for sowing by comparing the present climatic conditions with those of the previous year and the historical one, in addition to identifying the alerts that may affect production, as well as estimating the quantities of water necessary for irrigation, knowing the climatic factors that are affecting the crop and thus be able to generate a pest control based on a certain factor.

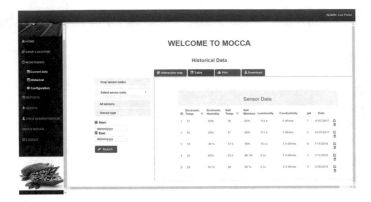

Fig. 8. Historical monitored data display from each sensor node.

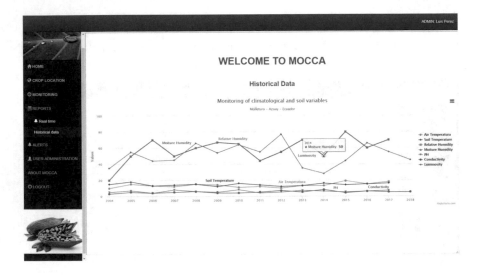

Fig. 9. Historical monitored data of climatological and soil variables.

5   Discussion and Conclusions

The presented design of an IoT architecture based on WSN applied to agricultural monitoring for cacao cultivation in Ecuador, constitutes an inclusive technological proposal for small and medium agricultural producers, a vulnerable sector and yet a priority in the country's economy. This project represents a low-cost technical solution that provides information on the main climatic variables and the physical and chemical parameters of a cacao plantation. The benefits of the MOCCA system lie in the use of the acquired data, since its manipulation leads to the improvement of three fundamental factors in a cultivation process: productivity, quality and sustainability.

The monitored information opens up the generation of preventive plans for the protection against pests and diseases, since certain climatic and soil moisture variables constitute ideal conditions for the proliferation of fungi, microorganisms, among others. The collected data can be used in agricultural techniques to safeguard the crop, keep the farmer aware of possible threats, and thus avoid losses.

The implementation of this tool guarantees the quality of the production since with the correct information of the state of the crop, specialized agricultural practices will be chosen, regulating the handling of agrochemicals. By improving quality, farmers will be able to satisfy the standards set for the marketing of their products.

The monitoring of agricultural variables in conjunction with a warning system is essential for the sustainable management of a crop because the proper management of resources depends on this information, resulting in an economically viable practice that satisfies production levels.

This proposal corresponds to an affordable system that monitors climatic data of temperature, humidity and luminosity, as well as soil moisture, pH and electrical conductivity through the use of open source and open hardware devices. It prioritizes in facilitating technology transfer to farmers through the design of a multiplatform application, with an intuitive and friendly interface where information is presented in the form of tables and statistical graphs that are easily understandable. Because of its ease of use the farmers would not depend on technical personnel for its continuous use, nor require a technical preparation for its manipulation. Another feature of the software is that it provides access to network information and monitoring history only to users registered and validated by the administrator.

In the future, an in-depth study will be carried out on issues related to communication and energy consumption. It is intended to expand the design with its respective implementation in coffee plantations.

Acknowledgement. The authors would like to express their gratitude to the support of the Thematic Network: RiegoNets (CYTED project 514RT0486), as well to PLATANO project by Telecommunications and Telematics Research Group (GITEL) from Universidad Politécnica Salesiana, Cuenca - Ecuador.

References

1. AGROCALIDAD, MAGAP: Manual de aplicabilidad de buenas prácticas agrícolas en cacao. AGROCALIDAD Agencia Ecuatoriana del Aseguramiento de la Calidad del Agro
2. Aiello, G., Giovino, I., Vallone, M., Catania, P., Argento, A.: A decision support system based on multisensor data fusion for sustainable greenhouse management. J. Cleaner Prod. **172**, 4057–4065 (2018). https://doi.org/10.1016/j.jclepro.2017.02.197
3. Altieri, M., Nicholls, C.: Agricultura tradicional y la conservación de la biodiversidad. Biodiversidad y uso de la tierra (1999)

4. ANECACAO Ecuador: Cacao: al vaivén del mercado. http://www.anecacao.com/es/noticias/cacao-al-vaiven-del-mercado.html
5. ANECACAO Ecuador: Cacao ccn 51. http://www.anecacao.com/es/quienes-somos/cacaoccn51.html
6. BCE: Exportaciones por producto principal (2018). https://contenido.bce.fin.ec/home1/estadisticas/bolmensual/IEMensual.jsp
7. Bongiovanni, R., Mantovani, E., Best, S., Roel, l.: Capítulo 1. Introducción a la agricultura de precisión. In: Agricultura de precisión: integrando conocimientos para una agricultura moderna y sustentable, pp. 13 – 22. Procisur/IICA, Montevideo, Uruguay (2006). www.iica.int
8. Cambra, C., Sendra, S., Lloret, J., Garcia, L.: An IoT service-oriented system for agriculture monitoring. In: 2017 IEEE International Conference on Communications (ICC), pp. 1–6, May 2017. https://doi.org/10.1109/ICC.2017.7996640
9. Carrión Santos, J.A.: Estudio de factibilidad para la producción y comercialización de cacao (Theobroma cacao L.) variedad CCN-51, Jama-Manabí. B.S. thesis, Quito (2012)
10. Ceballos, M.R., Gorricho, J.L., Palma Gamboa, O., Huerta, M.K., Rivas, D., Erazo Rodas, M.: Fuzzy system of irrigation applied to the growth of habanero pepper (capsicum chinense jacq.) under protected conditions in yucatan, mexico. Int. J. Distrib. Sens. Netw. 11(6), 123543 (2015)
11. Cevallos Barriga, J.M.: Producción y comercialización del cacao en el Ecuador período 2009–2010. B.S. thesis, Universidad de Guayaquil facultad de Ciencias Económicas (2011)
12. Culman, M., Portocarrero, J.M.T., Guerrero, C.D., Bayona, C., Torres, J.L., Farias, C.M.D.: PalmNET: an open-source wireless sensor network for oil palm plantations. In: 2017 IEEE 14th International Conference on Networking, Sensing and Control (ICNSC), pp. 783–788, May 2017. https://doi.org/10.1109/ICNSC.2017.8000190
13. Erazo, M., Rivas, D., Pérez, M., Galarza, O., Bautista, V., Huerta, M., Rojo, J.L.: Design and implementation of a wireless sensor network for rose greenhouses monitoring. In: 2015 6th International Conference on Automation, Robotics and Applications (ICARA), pp. 256–261. IEEE (2015)
14. Ferrández-Pastor, F.J., García-Chamizo, J.M., Nieto-Hidalgo, M., Mora-Pascual, J., Mora-Martínez, J.: Developing ubiquitous sensor network platform using internet of things: application in precision agriculture. Sensors 16(7), 1141 (2016). https://doi.org/10.3390/s16071141
15. Flores, K.O., Butaslac, I.M., Gonzales, J.E.M., Dumlao, S.M.G., Reyes, R.S.: Precision agriculture monitoring system using wireless sensor network and Raspberry Pi local server, pp. 3018–3021. IEEE, November 2016. https://doi.org/10.1109/TENCON.2016.7848600
16. Flores-Medina, M., Flores-García, F., Velasco-Martínez, V., González-Cervantes, G., Jurado-Zamarripa, F.: Monitoreo de humedad en suelo a través de red inalámbrica de sensores. Tecnología y ciencias del agua 6(5), 75–88 (2015)
17. García, S., Ramírez-Gallego, S., Luengo, J., Herrera, F.: Big data: Preprocesamiento y calidad de datos. novática 237, 17 (2016)
18. Grupo Banco Mundial: Agricultura, valor agregado. https://datos.bancomundial.org/indicador/NV.AGR.TOTL.ZS
19. Grupo Banco Mundial: Empleos en agricultura (2018). https://datos.bancomundial.org/indicador/SL.AGR.EMPL.ZS?locations=EC
20. Hernández Rojas, D.L., Mazón Olivo, B.E., Campoverde Marca, A.M.: Cloud computing para el internet de las cosas. caso de estudio orientado a la agricultura de precisión. I Congreso Internacional de Ciencia y Tecnología UTMACH (2015)

21. INEC: Estadísticas agropecuarias (2018). http://www.ecuadorencifras.gob.ec/estadisticas-agropecuarias-2/
22. Ipanaqué, W., Belupú, I., Castillo, J., Salazar, J.: Internet of things applied to monitoring fermentation process of cocoa at the piura's mountain range. In: 2017 CHILEAN Conference on Electrical, Electronics Engineering, Information and Communication Technologies (CHILECON), pp. 1–5. IEEE (2017)
23. Karimi, N., Arabhosseini, A., Karimi, M., Kianmehr, M.H.: Web-based monitoring system using Wireless Sensor Networks for traditional vineyards and grape drying buildings. Comput. Electron. Agric. **144**, 269–283 (2018). https://doi.org/10.1016/j.compag.2017.12.018
24. Kodali, R.K., Soratkal, S., Boppana, L.: WSN in coffee cultivation. In: 2016 International Conference on Computing, Communication and Automation (ICCCA), pp. 661–666, April 2016. https://doi.org/10.1109/CCAA.2016.7813804
25. Martínez Valle, L.: La Agricultura Familiar en el Ecuador. RIMISP (2013)
26. Ministerio de Turismo: El cacao como producto símbolo del ecuador", la meta de la cacao cultura (2018). https://www.turismo.gob.ec/el-cacao-como-producto-simbolo-del-ecuador-la-meta-de-la-cacao-cultura/
27. Morales Intriago, F.L.: Los productores de cacao tipo nacional en la provincia de los ríos-ecuador: un análisis socioeconómico (2013)
28. Núñez Cárdenas, F.D.J.: El proceso de minería de datos. Ciencia Huasteca Boletín Científico de la Escuela Superior de Huejutla **1**(1) (2013)
29. Ojha, T., Misra, S., Raghuwanshi, N.S.: Wireless sensor networks for agriculture: the state-of-the-art in practice and future challenges. Comput. Electron. Agric. **118**, 66–84 (2015). https://doi.org/10.1016/j.compag.2015.08.011
30. de la Piedra, A., Benitez-Capistros, F., Dominguez, F., Touhafi, A.: Wireless sensor networks for environmental research: a survey on limitations and challenges. In: 2013 IEEE EUROCON, pp. 267–274. IEEE, July 2013. https://doi.org/10.1109/EUROCON.2013.6624996
31. Scholz, B.I., Morales Rodríguez, J.A., Mena Giacometti, J.R., Aguilar Meneses, P., Reinoso Valarezo, A., Cando Cando, P.: Informe Nacional del Ecuador para la Tercera Conferencia de las Naciones Unidas sobre Vivienda y Desarrollo Urbano Sostenible HABITAT III. Ministerio de Desarrollo Urbano y Vivienda del Gobierno de la República del Ecuador (2015)
32. W3Techs - World Wide Web Technology Surveys: Usage of server-side programming languages for websites (2018). https://w3techs.com
33. Wixted, A.J., Kinnaird, P., Larijani, H., Tait, A., Ahmadinia, A., Strachan, N.: Evaluation of lora and lorawan for wireless sensor networks. In: 2016 IEEE SENSORS, pp. 1–3. IEEE (2016)

Silkworm Growth Monitoring in Second Stage -Instar- Using Artificial Vision Techniques

Luis Javier Suárez[1], Yaneth Patricia López[1],
Wilfred Fabian Rivera[1(✉)], and Agapito Ledezma[2]

[1] Technology Development Center, Corporation Cluster CreaTIC,
Popayán, Colombia
ceo@clustercreatic.com
[2] University Carlos III of Madrid, Madrid, Spain

Abstract. In the Department of Cauca (Colombia), there is evidence of low efficiency in the process of generating silk cocoons by small and medium producers due to the manual monitoring of worms natural growth; traditionally, a person without an advanced technical knowledge but with empirical experience has evaluated the appropriate feeding time based on the concentration of worms. This task becomes more expensive and inefficient when the production of worms increases. For this reason, we propose to improve the aforementioned process through the analysis of worm bed images in its second stage –instar-, by automatically determining the most suitable period for feeding using artificial intelligence techniques. The experiments showed promising results that will guide automation at low costs in the worm breeding industry for this region.

Keywords: Artificial vision · Machine learning · Monitoring · Silkworm

1 Introduction

Sericulture is an agro-industrial activity mainly developed in rural or suburban areas, this activity has historically been considered as a complement of agricultural production systems. It benefits small and medium producers who work as a family nucleus around silk production using their own local techniques. Although the activity does not need significant initial investment, it does require farmers to provide controlled climatic and sanitary conditions during the worm breeding and feeding program which is based on quality mulberry [1]. Silk has been used for 5,000 years and it is known as the queen of textiles due to its brightness, softness, elegance, durability, and traction properties [2]. It has good resistance and absorption properties, making it ideal for both, hot and cold climates. From its origin, garments derived from this fiber have been a symbol of status and luxury, it is said that by 1,300 BC, when it was discovered by the Chinese, this fiber was reserved only for members of the royal family and through the years, its usage expanded to the rest of Asia and, later, to Europe and Latin America during the time of the conquest [3]. Sericulture as a mean to obtain fiber and silk thread has an economic relevance in approximately 40 countries. China leads with 84% of the production of raw silk marketed worldwide, followed by India with 15%. These two countries together produced 202,072 tons in 2015. The sericulture activity in the

© Springer Nature Switzerland AG 2019
J. C. Corrales et al. (Eds.): AACC 2018, AISC 893, pp. 58–72, 2019.
https://doi.org/10.1007/978-3-030-04447-3_4

Department of Cauca in Colombia is carried out in rural areas by producers and artisans who receive an income for the production of fresh buds, yarns, fabrics and/or garments.

The silkworm is a completely domesticated insect used for the production of silk in countries located in the tropical strip. It needs appropriate conditions to develop and grow during the larval cycle (40–50 days approximately). This duration will depend on weather conditions provided in the breeding sheds; with special attention in temperature and relative humidity. It is essential to determine the specific moment to supply food for each instar (stage), as well as the beginning and ending of the molting stage, since these decisions will directly affect the production of cocoons, in terms of quantity and quality.

This research was developed with the support of The Corporation for the Development of the Sericulture of Cauca –CORSEDA-, which is an organization that was born in 2001, but its activities began around 1998. It started with 200 families producing cocoons and silk handicrafts in the center and north area of the Department of Cauca, Colombia.

CORSEDA is a diverse organization, it is the result of the ethnic and cultural diversity of the families that make up the Departament of Cauca: afrodescendants, indigenous people (Paez and Guambia ethnic groups), and farmers that work together for a single common purpose: To improve the standard of living of their families and to offer to the world an inimitable product that incarnates the sum of different perceptions about the environment and life.

Currently in the Department of Cauca, the feeding time for the breeding sheds is made by observation and/or qualitative evaluation, a method that shows flaws expressed by non-uniform larvae populations in each instar, which are altering the breeding lot. Similarly, every decision about the first instar will influence unequivocally in the following ones and the probabilities of recovering the genetic advantage of the hybrid that is raised is minimal. In this sense, the greatest difficulty for decision making occurs in the first instars of the silkworm cycle, since the approximate larval size is between 3 and 12 mm; according to [4] newborn larvae with a length of 3 mm, those of first instar of 7 mm and those of second 12 mm. Taking into account the larval production strategy of the Corporation for the Development of the Sericulture of Cauca (CORSEDA), it is relevant to select the first instar to support and strengthen decision making, since the Corporation has a person who incubates and provides first-day larvae of third instar to all associated sericulturists, in this sense the quality and uniformity of the larvae to be delivered is fundamental.

The research proposes a method of semantic segmentation of images as an assistance system in silkworm feeding processes. This system incorporates the implementation of sensors in a dynamic infrastructure, which is not very solid, and a strategy that supports the decision-making process of the sericulturists during feeding periods. In Sect. 2, we present the works related to the production of silkworms, as well as the different approaches used for the semantic segmentation of the images. In Sect. 3, the domain is described. The proposed approach is presented in Sect. 4 and the experimental configuration is detailed in Sect. 5. Finally, Sect. 6 describes the conclusions, which show the usefulness of this proposal and its potential impact on future development.

2 Related Works

Image segmentation is a widely studied problem, which commonly serves as a basis for higher-level tasks in image and video processing. In 2010 there were five categories of image segmentation algorithms [5, 6]: thresholding [7, 8], template matching [9, 10], clustering [11–13], edge detection [14–16] and region growth [17, 18]. These algorithms have proven to be successful in many applications, however, [19] and its group lists a series of important problems that occur when these algorithms are applied to more complex images, larger in size or with many edges. Image thresholding methods are popular due to their simplicity and efficiency. However, thresholding methods cannot separate those areas which have the same gray level. Semantic image segmentation attempts to divide an image and classify each of these parts into a series of predetermined classes. In other works, the author tries to classify each of the pixels of the input image within one of the previously defined classes [20–23]. This is known as pixel-wise semantic segmentation and generates an image as a result in which each pixel has the numerical value of the class in which it was labeled by the algorithm. For this purpose, some publications define an automatic learning model whose output will be the value of each pixel of the resulting image and its input corresponds to the color characteristics of the corresponding pixels in the input image [19]. Also, a conventional watershed segmentation algorithm can be used, it is a popular method in region-based partition technique. However, it is not suitable for color image directly because it is difficult to determine color intensity [24]. When precise limits of the object are not required, for example, to count the number of silkworm eggs, the object detection technique based on the position of the centroid that is suitable for elliptical objects can be used. Subsequently, the color image becomes a fine binary image and the foreground pixel is separated from the background by the adaptive thresholding technique [24]. Other image processing techniques focus on monitoring the silkworm growth rate. For this, the average length and average width of silkworm larvae were automatically measured in real time. Moreover, area of green castor leaves was also measured in order to monitor an amount of food supply. The above was achieved by sending the video sequence of the breeding bed to analyze in a computer, using an algorithm that measured green leaves area choosing only the green plane of an RGB color image, appropriate adaptive limits and after eliminating the noise a binary image is obtained from which it is possible to extract the required area. From the experiment, the decreasing area resulted in a histogram change by time, therefore an adaptive thresholding was necessary to separate the silk worms, the green area and the background [25].

For the current trends there are works that use internet of things technology with a focus based on 6LoWPAN (IPv6 over Low power Wireless Personal Area Network) to design a real-time monitoring and disinfection actuating system for sericulture with an image processing technology to identify the stages of silk worm life cycle. In this work, the prototype supports real time data collection, being designed and implemented using Contiki OS in order to control the atmospheric condition inside the system according to the requirements in the stages of sericulture life cycle. This prototype is based on CoAP (Constrained Application Protocol), RPL (Routing Protocol for Low power and lossy Networks) and 6LoWPAN protocols.

In order to auto capture the pictures and analyzing them, the prototype was developed using the TelosB motes that runs 6LoWPAN stack interfaced to temperature and humidity sensors with an disinfection actuation system and a serial camera. For the above mentioned, it was able to check the status on sericulture process.

The image processing is used to identify the colour change in the body of the worms, which indicates different stages [26].

3 The Dominion

3.1 Sericulture in the Department of Cauca

Sericulture in the Department of Cauca is still artisanal which implies, among other things, that the type of materials used for the construction of the breeding sheds are adjusted to the economic capacity of the producer and, generally, do not have the adequate conditions for the management of temperature and relative humidity [27]. The manufacture of the breeding areas includes the use of brick baked from ground, brick-based guadua or simply transparent plastic for the walls; the floors of concrete or ground and the roof in zinc tiles or cement, the ceiling in wooden boards or guadua mat and the windows covered with transparent plastic to allow the entrance of light and sackcloth to favor an adequate ventilation.

3.2 Life Cycle of the Silkworm

The life cycle of the silkworm consists of five stages or *instars*; Depending on the stage the worm is in, its feeding will vary. In the Department of Cauca, cocoon production begins with the incubation of the Pilamo I hybrid eggs. For this, CORSEDA uses an incubator designed to guarantee the maintenance of temperature and relative humidity. The way to feed the newborn larvae, and 1st and 2nd instar is by finely chopped mulberry (1–2 cm^2) with apical leaves, whose texture and nutrient content is appropriate for these stages. Once the 2nd moult ends and the larvae enter the third instar, the breeding lot is delivered to the producers.

In the breeding sheds, the sericulturists use different alternatives to control the temperature and relative humidity. They are used from gas stoves to charcoal in a clay pot, which entails certain limitations when it comes to regulating and/or maintaining the temperature. In Fig. 1, it can be seen the status of the larvae in 3rd instar in the breeding sheds of the sericulturist.

During instar 3, 4 and 5, the sericulturist uses mulberry branches in the three rations that are supplied per day to the worms; For this reason, the mulberry pruning is done above the production head and the rest of the plant is used. It should be noted that the amount of rations per day also depends on the sericulturist and the observation and analysis carried out in the cabins. For example, if the larvae consume the mulberry quickly, some producers make the decision to provide one or two additional servings per day to the worms.

Fig. 1. Stage of the larvae in the 3rd instar in the breeding sheds of the sericulturists.

On the other hand, the cabins used for breeding worm larvae are usually constructed with guadua. Regarding the number of cabins owned by each producer, this will depend on the breeding capacity they have, in other words, the availability of mulberry. 80% of the farmers produce ¼ of a box (between 5,000–6,000 larvae).

In the last instar of the biological cycle of the silkworm, -5th instar- the producer must have available the bodies of rods -rodalinas- for the larvae (See Fig. 2). The stage begins, generally, on the sixth day of the 5th instar and it lasts three days as long as suitable conditions of temperature and relative humidity are maintained and the nutritional requirements of the larvae and the correct handling of the molt have been met. In this instar, the feeding of the larvae plays a fundamental role given that it will influence the weight of the cocoon bark. In addition, a temperature between 24–25 °C must be maintained during the cocoon formation process, if the temperature decreases, the bark will not form properly and it will be soft, affecting the qualification received by the sericulturist in the analysis of quality that CORSEDA carries out.

4 Proposed Approach

Proposed approach in this paper is oriented to differentiate images from worm's bed, larvaes and mulberry leaf, analyzing periodically variations in the relation mulberry-worm, enriching such relation with movements information from larvaes, temperature and relative humidity of environment. Thus, the approximation proposed consists of fourth main phases, (a) first it gets new representations from the original image based on spatial filters and changes of color spaces; (b) samples are extracted pixel by pixel for all the images obtained, labeling every sample in every wanted class (mainly: worm and mulberry), and then, they can be used as a set of data and starting from them, (c) it

Fig. 2. Bodies of rods -rodalinas- for the larvae.

is used an automatic learning algorithm to obtain the segmented image; (d) additionally, this procedure is complemented with a process of backgroud removal, which is the elimination of all information that does not include worms and leaves mulberry (See Fig. 3).

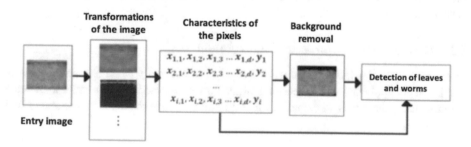

Fig. 3. Proposal approach.

4.1 Image Transformation

First at all, spatial filters are applied to images. The resulting images from the smoothing filters contain information about their corresponding neighbors in every pixel. i.e. they consider the context surround to every pixel. Also, it is applied a bilateral filter, which is able to remove noise, besides holding the sharp edges. This filter is especially useful to highlight the worms. Additionally, it is used a representation in gray scale, including histogram equalization of color. finally, it is included a filter of smoothing strong based on median, what results especially useful to know the illumination in every area of the worm's bed. The parameters of every transformation have been empirically adjusted.

All transformations and their functions and parameters in OpenCV are shown in Table 1.

Table 1. Transformation and parameters.

Transformation of the image	Channels	Function Cv2	Parameters
Normal image	h, s, v	–	–
Bilateral filter	h, s, v	bilateralFilter	d = 9, sigmaColor = 75, sigmaSpace = 75
Gaussian blur image	h, s, v	GaussianBlur	ksize = (5, 5)
Color equalized image	h, s, v	equalizeHist	–
Median blur image	h, s, v	medianBlur	ksize = 201
Gray scale image	gray	cvtColor	–
Gray scale equalized image	gray	equalizeHist	–

4.2   Feature Generation

Once all different transformations about the images were realized, a total of 17 features were generated. Every transformation of image in color contributed 3 values to the features vector to every pixel, while images in grayscale contributed just one feature. Finally, in Fig. 4 it can be observed the relevance of the features with respect to the classification task.

4.3   Classification of Worms and Background

In order to classify pixels in worms or leaves, it is necessary to label manually every pixel and to apply automatic learning algorithm supervised. In this case in labeled images phase, there was an expert in the field and an image database management tool (See Fig. 5).

4.4   Background Removal

Finally, it is necessary to remove the part of the image that corresponds to the background of worms' bed. The bed of the silk worms has particularities in the background, which are given by a blue background, gray trellis and soil. The lattice has blue, red and violet colors.

Performing the analysis of the relationship between the number of worms and the amount of food is required, therefore, the background removal process of the original image consists of the following steps (See Fig. 6):

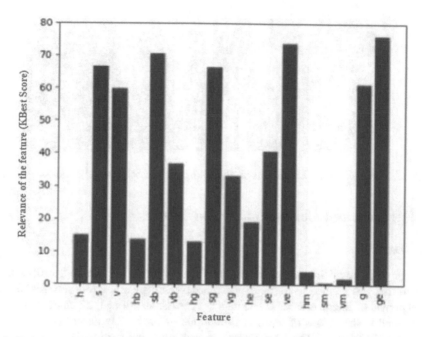

Fig. 4. Relevance of characteristics. (h,s,v: normal image. hb, sb, vb: bilateral filter. hg, sg, vg: gaussian filter.

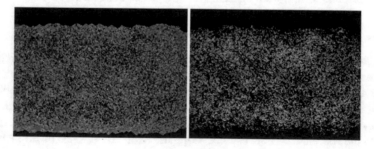

Fig. 5. Images resulting from the classification process. Left: Image corresponding to food (leaves). Right: Image corresponding to worms.

a. Removal of colored marks (or flags) located on the trellis of the bed
b. Removal of blue background paper
c. Removal of leaves according to a classifier between leaves and background
d. Removal of worms according to a sorting algorithm between worms and background
e. Morphological closing and opening operations
f. Final result.

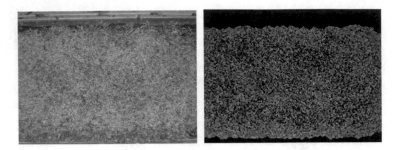

Fig. 6. *Left*: Original image. *Right:* Image with background removal.

5   Experimental Configuration and Results

5.1   Data Set

In order to distinguish between worms and leaves in the growth bed, a data set was created that was labeled from the expert's observations using a tool developed for the management of image data sets. The data set has, as mentioned, 17 characteristics that correspond to the values of the transformations of previously described images for a specific pixel location. The label (class) of each pixel is worm or leaf.

A total of 958 samples were collected, taken from the worms bed in their second instar, of which 417 were labeled as pixels of worms and 541 as leaves. The original data set was divided into training data (75%) and test data (25%).

5.2   Learning Algorithms

In order to validate different learning algorithms, the performance and prediction time for a given amount of data were taken into account. The classification time is important since the model will be implemented in a device with limited computing capabilities. In this way, it is measured how long each model trained in making the prediction of the complete data set takes. Based on tests conducted on a Raspberry Pi, the total time used to classify the complete data set can not be greater than 0.04 s, since its implementation in hardware would be unfeasible in the proposed system.

The measurement of the overall accuracy chosen for the evaluation of the models is based on [28] where the authors use the measurements of F Score, precision and recall for the accuracy evaluation, additionally in [24] the measurement of F Score is used as the only measure of performance. In this paper, the evaluation of general accuracy of each model is made with the measurement of F Score that takes the measurement of precision and recall into account. The algorithms logistic regression, Linear Vector Support Machines, Random Forest, decision tree, Multilayer Perceptron and k-NN were used. For each algorithm, different learning parameters were used (see Table 2).

Table 2. Algorithms and learning parameters.

Model	Hyperparameters
AdaBoost classifier	n_estimators = [17, 22, 27, 32, 37, 42]
Logistic regression	C = [1, 1e2, 1e3, 1e4, 1e5, 1e6, 1e7, 1e8, 1e10, 1e20]
Random forest	max_depth = [3, 4, 5, 6, 7, 8, 9], n_estimators = [3, 4, 5, 6, 7]
Decision tree	criterion = ['gini','entropy'], max_depth = [4, 5, 6, 7, 8, 9, 10, 11, 12, 15, 20, 30, 40, 50, 70, 90, 120, 150]
Linear SVC	C = [1e–5, 1e–4, 1e–3, 1e–2, 1e–1]
K-neighbors classifier	n_neighbors = [1, 2, 3, 4, 5, 6, 7], leaf_size = [10, 30]
MLP classifier	Alpha = [0.3, 0.4, 0.5, 0.6], hidden_layer_sizes = [300, 400, 500, 600]

5.3   Model Selection

Figure 7 shows the results of the two models generated of greater relevance and better results according to the evaluation criteria (classification time, F value): (a) a model that obtains a very high F value (0.9918 for k-NN) but consume a high time in the classification, so it does not meet the time requirement of prediction; (b) the model obtains an adequate F value (0.9762 in the decision tree) and its prediction times are within the desired range. The latter model has the highest F value, which corresponds to a *decision tree* with the learning parameters "*max_depth = 50*" and "*criterion = entropy*". Finally, the selected model is trained with the training set and evaluated in the test set. With this way, an F value of 0.966 is obtained using the test set.

5.4   Results

From the segmentation of the image using the selected model it is possible to obtain labels for each segment, in particular for food, worms and bottom. This process is done by reading each pixel of the image. Based on this, the following image features, called low-level features, are extracted:

- *leafs_px_count:* Pixel count of leaves.
- *worms_px_count:* Pixel count of worms.
- *total_px_amount:* Sum of worm and leaf pixel count (Excluding background): *total_px_amount = leafs_px_count + worms_px_count*
- *worms_avg_h, worms_avg_s, worms_avg_v:* average color (h, s, v) of all the pixels detected as worms.
- *leafs_avg_h, leafs_avg_s, leafs_avg_v:* average color (h, s, v) of all the pixels detected as mulberry leaves.

The counting characteristics of pixels corresponding to worms and of pixels corresponding to food (mulberry) are obtained from the segmented image. These characteristics are particularly useful because they show the typical behavior of the silk worms, which are completely covered with mulberry leaves in each feeding and then

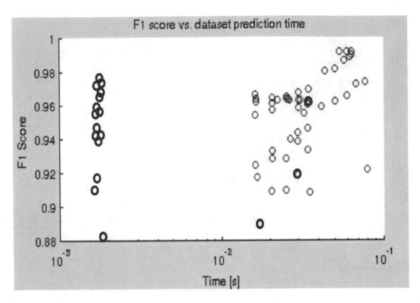

Fig. 7. Selection criteria for different models and values of learning parameters. (Criteria: F value and prediction time).

they begin to ascend progressively in search of more food. This behavior can be inferred from Fig. 8, where four feeds are observed (when the count of leaves is equal to the total of pixels), and where it is observed that from the beginning of each feeding there is a gradual increase in the count of pixels of worms.

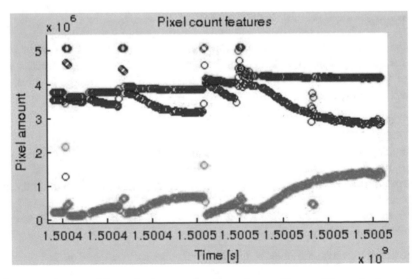

Fig. 8. Image features: Blue - pixel count of mulberry leaves. Green - Worm pixel count. Red - Total pixel count with background removal.

Another feature with important information is the average grayscale (Channel V) of the pixels classified as worms. In Fig. 9 it is observed that this characteristic tends to increase throughout the experiment, being possible to relate it to the age of the worms. This occurs because the type of worms in this region is gray characteristic in its early stages and becomes clearer and clearer as its development progresses.

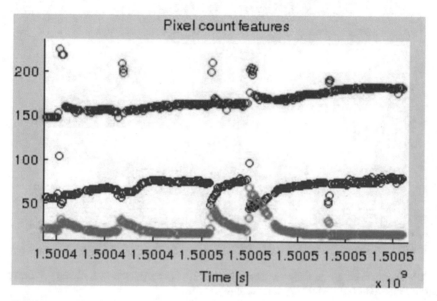

Fig. 9. Average of HSV channels for pixels classified as worms: Blue - Channel H. Green - Channel S. Red - Channel V.

From the above mentioned characteristics, it is possible to obtain other relevant ones, called high level features, since they are generated from low level ones, and are:

- *leaf_worm_rate*: mulberry ratio (food) to worms.

  leaf_worm_rate = 100(leafs_px_count/worms_px_count)*

- *leaf_px_count_norm*: pixel count of mulberry leaves normalized with respect to the total of pixels with background removal.

  leaf_px_count_norm = leaf_px_count/total_px_count

- *worm_px_count_norm*: Worm pixel count normalized with respect to total pixels with background removal.

  worm_px_count_norm = worm_px_count/total_px_count

Figure 10 shows the variation of *high-level features* over time, which provides relevant information about the behavior of the worm during the second instar. The points of inflection in each curve show a significant variation, directly related to feeding moments. Thus, according to the blue color curve, when the worm is fed, there is an increase in the pixel count of leaves; to the extent that the worms feed, this curve presents a decreasing tendency, which is recovered once the farmer feeds the worms again. On the other hand, the curve of green color, shows a growing trend, an aspect that indicates the growth of the worm during this stage of its life cycle.

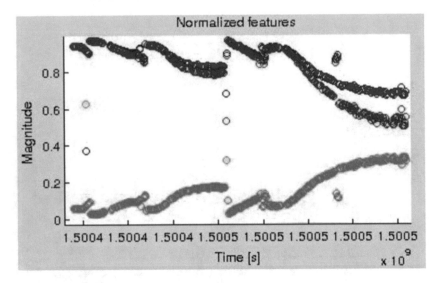

Fig. 10. Normalized image features. Blue - Normalized sheet pixel count. Green - Worm pixel count normalized. Red - Mulberry relationship to worms.

6   Conclusions

This paper proposes a method of semantic segmentation of images as a component of a system of assistance in silkworm feeding processes for small and medium producers in the Department of Cauca (Colombia). This system contemplates the deployment of sensors in a flexible infrastructure, forming a strategy that supports the decision-making process of the sericulturist in terms of feeding periods in the cultivation areas.

The results of the generated segmentation model evidence the relevance of the approach proposed for the automatic monitoring of the silkworm in the second instar of its life cycle, through the analysis of images of the bed of worms. With this, we move towards a system that supports the producer of the sericulture chain in estimating the most suitable period for feeding the worms based on artificial intelligence, allowing them to increase their efficiency and productive quality in this industry.

7   Future Work

As future work, it is possible to emphasize performance measures using tools such as the confusion matrix that allows generating new findings for the application of each model and selecting models with better precision.

On the other hand, it is possible to work on the improvement of the classification model of silkworms and leaves, since this allows to optimize the responses so the behavior of silkworms in the second instar can be determined with greater precision.

In the same way, it is possible to work in a real-time analysis module for the characteristics of the current periodicity of the images and the recorder: the movement of the worms, the change of the deepest relationship and the growth in the second instar; as well as, context variables: temperature and relative humidity.

The above information represents the behavior during its second instar according to its context variables, with which a model can be generated to predict the appropriate time of its feeding and thus guarantee a growth of the greatest uniformity during this stage of its cycle lifetime.

Once the previous module is available, it is possible to generate an alarm mechanism, alerting those responsible for the care of the worms in order to feed them.

Acknowledgment. The authors acknowledge the company Quasar Tech SAS, which generated the necessary software and hardware to implement the monitoring system of the silkworm bed in its second instar and the Corporation for the Development of the Sericulture of Cauca (COR-SEDA), which facilitated the space and qualified personnel to deploy the experiments carried out within the framework of the project "STRENGTHENING OF THE MARKET VERTICAL IN AGRIBUSINESS IN COMPANIES OF CLUSTER CREATIC", financed by Colombian agencies as: MinTIC, COLCIENCIAS and the Government of Cauca.

References

1. Pescio, F., Zunini, H., Basso, C.P., de Sesar, M., Frank, R.G., Pelicano, A.E., Vieites, C.M.: "Sericicultura: manual para la producción", Buenos Aires, Inst. Nac. Tecnol. Ind. e Univ. Buenos Aires-Facultad Agron

2. Madihalli, P.D.B., Ittannavar, P.S.S.: Arduino based automated sericulture system **14**(1) (2017)

3. Maya, A., Cardona, J.E., Álvarez, A., Elena, B., Arenas, A., González, D., Zapata Sierra, I., et al.: Aplicación de la Norma Gots a un proceso productivo: caso Corseda (2013)

4. Cifuentes Correa, C.A., Sohn, K.W.: Manual técnico de sericultura: cultivo de la morera y cria del gusano de seda en el trópico., no. Doc. 21309 CO-BAC, Bogotá (1998)

5. Unnikrishnan, R., Pantofaru, C., Hebert, M.: Toward objective evaluation of image segmentation algorithms. IEEE Trans. Pattern Anal. Mach. Intell. **29**(6), 929–944 (2007)

6. Estrada, F.J., Jepson, A.D.: Benchmarking image segmentation algorithms. Int. J. Comput. Vis. **85**(2), 167–181 (2009)

7. Maitra, M., Chatterjee, A.: A hybrid cooperative–comprehensive learning based PSO algorithm for image segmentation using multilevel thresholding. Expert Syst. Appl. **34**(2), 1341–1350 (2008)

8. Yüksel, M.E., Borlu, M.: Accurate segmentation of dermoscopic images by image thresholding based on type-2 fuzzy logic. IEEE Trans. Fuzzy Syst. **17**(4), 976–982 (2009)

9. Zeng, X.-Y., Chen, Y.-W., Nakao, Z., Lu, H.: Texture representation based on pattern map. Signal Process. **84**(3), 589–599 (2004)

10. Of, O.: Template-based automatic segmentation of drosophila mushroom bodies **113**, 99–113 (2008)

11. Wang, X.-Y., Sun, Y.-F.: A color-and texture-based image segmentation algorithm. Mach. Graph. Vis. Int. J. **19**(1), 3–18 (2010)
12. He, R., Datta, S., Sajja, B.R., Narayana, P.A.: Generalized fuzzy clustering for segmentation of multi-spectral magnetic resonance images. Comput. Med. Imaging Graph. **32**(5), 353–366 (2008)
13. Shi, J., Malik, J.: Normalized cuts and image segmentation. IEEE Trans. Pattern Anal. Mach. Intell. **22**(8), 888–905 (2000)
14. Bao, P., Zhang, L., Wu, X.: Canny edge detection enhancement by scale multiplication where **27**(9), 1485–1490 (2005)
15. Christoudias, C.M., Georgescu, B., Meer, P.: Synergism in Low Level Vision, no. 1
16. Chung, K., Yang, W., Yan, W.: Efficient edge-preserving algorithm for color contrast enhancement with application to color image segmentation **19**, 299–310 (2008)
17. Enrique, L., Ugarriza, G., Saber, E., Amuso, V.J.: Automatic image segmentation by dynamic region growth and multiresolution merging by (2007)
18. Peter, Z., Bousson, V., Bergot, C., Peyrin, F.: A constrained region growing approach based on watershed for the segmentation of low contrast structures in bone micro-CT images **41**, 2358–2368 (2008)
19. Wang, X.-Y., Wang, T., Bu, J.: Color image segmentation using pixel wise support vector machine classification. Pattern Recognit. **44**(4), 777–787 (2011)
20. Long, J., Shelhamer, E., Darrell, T.: Fully Convolutional Networks for Semantic Segmentation
21. Bittel, S., Kaiser, V., Teichmann, M., Thoma, M.: Pixel-wise Segmentation of Street with Neural Networks, pp. 1–7 (2010)
22. Wu, C., Cheng, H., Li, S., Li, H., Chen, Y.: ApesNet : A Pixel-wise Efficient Segmentation Network (Invited Special Session Paper) (2016)
23. Paisitkriangkrai, S., Sherrah, J., Janney, P., Hengel, A.V.: Effective Semantic Pixel labelling with Convolutional Networks and Conditional Random Fields
24. Kiratiratanapruk, K., Watcharapinchai, N., Methasate, I., Sinthupinyo, W.: Silkworm eggs detection and classification using image analysis. In: 2014 International Computer Science and Engineering Con-ference (ICSEC), pp. 340–345 (2014)
25. Leelertyanon, I., Areekul, V.: The automatic measurement of silkworm growth rate and leaf's area using image processing. In: 2002 IEEE International Conference on Industrial Technology IEEE ICIT 2002, vol. 1, pp. 242–245 (2002)
26. Adarsh, U., Shivayogappa, H.J., Navya, K.N., et al.: Automated smart sericulture system based on 6LoWPAN and image processing technique. In: 2016 International Conference on Computer Communication and Informatics (ICCCI), pp. 1–6 (2016)
27. Grisales-Muñoz, C.: Caracterización de tres unidades serícolas en los municipios de Piendamó y Morales, Cauca, (Tesis de pregrado), Universidad del Cauca, Popayán (2015)
28. Kiratiratanapruk, K., Sinthupinyo, W.: Silkworm egg image analysis using different color information for improving quality inspection. IEEE (2016)

Affectations in Soil Fertility in the Direct Influence Area Downstream of the Grande River: Chone Multiple Purpose Dam in Ecuador

Oswaldo Borja-Goyes[1]([✉]), David Carrera-Villacrés[2],
David González-Riera[1], Edgar Guerrón-Varela[1],
Paula Montalvo-Alvarado[1], and Andrés Moreno-Chauca[1]

[1] Universidad Central del Ecuador, Quito, Facultad de Ingeniería en Geología
Minas Petróleos y Ambiental (FIGEMPA), Quito, Ecuador
{oaborja,degonzalez,pjmontalvo,asmoreno}@uce.edu.ec,
erguerron@espe.edu.ec
[2] Universidad de las Fuerzas Armadas – ESPE, Departamento de Ciencias de la
Tierra y la Construcción, Sangolquí, Ecuador
dvcarrera@espe.edu.ec
www.gica.espe.edu.ec

Abstract. The agricultural activity in Chone is of utmost importance to its inhabitants, it is one of the main economic activities in the area, which is characterized by the production of cultivated pasture, cocoa, banana, corn and other agricultural products that use soil as the main resource. This study determined the evolution of the physical and chemical parameters (pH, electric conductivity, calcium, magnesium, potassium, and sodium) related to the fertility of the soil, downstream from the dam of Grande River, Multiple Purpose Project Chone (direct influence area). The Analysis of data belonging to nine sampling points, was performed by software R, from the linear regression and testing of waste to verify the reliability of the data, this analysis allowed to determine that between 2013 and 2016, in all the ions of study, there is a loss or decline in the initial concentration (year 2013) with respect to the final concentration (year 2016), as in the case of potassium, which presents an average loss of 94.88% up to the year 2016. Additionally, the tendency of the data is increased, indicating that with the passage of time, this will increase, largely affecting the productivity of the soil in the study area.

Keywords: Agricultural · Calcium · Magnesium · Ion · Influence Sodium

Grupo de Investigación en Contaminación Ambiental (GICA).

J. C. Corrales et al. (Eds.): AACC 2018, AISC 893, pp. 73–88, 2019.
https://doi.org/10.1007/978-3-030-04447-3_5

1   Introduction

1.1   The Soil and Its Fertility

The soil, as a physical support and source of nutrients, is a fundamental resource for the proper development and good production of any plant species (Andrades and Martínez 2014). In the evaluation of soil fertility, its chemical, physical and biological properties are considered (García, et al. 2012). The most relevant parameters that intervene in soil fertility are: texture, pH, electrical conductivity (EC), organic matter (OM), phosphorus (P), potassium (K), total change capacity, total carbonates and active limestone (Andrades and Martínez 2014).

1.2   Effects on Soils by the Construction of Dams

With the construction of a dam, in ecological terms, there is a drastic change in the use of the land, with a considerable extension of the vegetation cover and the riparian ecosystems altered by the flooding area of the reservoir. It is known that the impacts downstream are of greater scope than those upstream, by interrupting or modifying the processes of erosion and deposition of sediments, as well as the natural flooding of the plains and the recharge of water from the subsoil.

Any change in the sediment balance influences the distribution of vegetation in the vicinity of the river (Yrízar et al. 2012), this lack of sediment, is due to the fact that the water, which contributes the sediments as a natural process, leaves the dams leaving practically all their sedimentary load (García and Puigdefábregas 1985). Because of result of soil degradation after losing fertility, there is a shortage of water and the soil loses the natural flood management capacity (Arroyo 2016).

1.3   Soil Degradation, Infertility and Toxicity

According to FAO World Soil Chart (FAO 2015), soil degradation is understood as the partial or total loss of its productivity, quantitatively or qualitatively, as a consequence of processes such as hydraulic and wind erosion of soils, salinization, waterlogging, depletion of plant nutrients, deterioration of soil structure, desertification and pollution. Soil degradation affects agricultural productivity, aggravated by floods, landslides and accumulation of sediments in rivers.

Physical degradation of soil affects its structure, decreases permeability and water retention capacity. Chemical degradation occurs when there is an affectation of the organic and inorganic species of the soil, with a reduction of the nutrients available for the vegetation, which leads to a gradual acidification of the soil and an increase in its toxicity, favoring polluting substances mobility (Cárdenas and Erazo 2013).

1.4   Land Use and Economy of Chone

Soil conservation is of great importance for the economy of Chone, which according to the population and housing census of 2010, 38% of the economically active population of the canton is engaged in agricultural activities (GAD Chone, 2014), for this reason soil conservation is vital for the economic activities of the Chone population, it is also important for the conservation of native plant species, which cover 14.18% of Chone's total area (Instituto Espacial Ecuatoriano - IEE 2013).

Soil productivity in this area was threatened by periods of little water and heavy rainfall, which in both cases caused the loss of crops, for this reason, the State built the Chone Multiple Purpose Project (PPMCH) with the objective of avoiding floods and meet the demand for irrigation water (Carrera-Villacrés, et al. 2015). This project is in operation since November 24, 2015, when it was inaugurated (Empresa Pública del Agua 2015).

The objective of the present work was to analyze the evolution of the physical and chemical properties of the soil related to its fertility, in the area of direct influence of the PPMCH, downstream of the Grande River dam, in the years 2013 and 2016. Determine if there are deficiencies in fertility, identifying the existing correlation between the data obtained and a residue test to verify the reliability of the same, through the software R.

1.5   Ion Deficiency for the Cultivation of Cacao

A soil without required nutrients have a low yield, in the plants it increases the susceptibility to pests and diseases. The deficiency of potassium ion appears initially in the oldest leaves, the leaves of the buds and suckers are getting smaller as the deficiency is accentuated. The central area of the leaf becomes necrotic in the case of severe deficiencies. In cases of severe calcium ion deficiency, premature leaf fall and shoot death occurs. In the older leaves the apical and marginal burning progresses rapidly. Calcium deficiency causes decreased root growth. The deficiency of Mg appears as a chlorosis that begins in the areas near the central vein of the oldest leaves and after a time the symptom diffuses between the veins towards the edges of the leaf (Vele 2013).

1.6   Ion Deficiency for the Cultivation of Corn

The primary elements such as Nitrogen, Phosphorus and Potassium are determinants for the growth of the plant, normally the soil cannot supply it in sufficient quantities for the normal development of the plants therefore it needs to be incorporated in the form of synthetic fertilizers or fertilizers. Potassium deficiency is closely related to the production of sugars, cellulose, proteins and starches, regulating the respiration of cells (Garcia 2013).

2   Material and Methods

2.1   Materials

The study area is located downstream of the Río Grande dam, in the area of direct influence of the PPMCH, specifically in the vicinity of the city of Chone, cantonal capital of Canton Chone, province of Manabí (Ecuadorian coast), west of the Republic of Ecuador. According to INAMHI and FAO, the city is located at coordinates 601512 E,

9923462 N WGS84 UTM Zone 17S (INAMHI; FAO; 2008). Figure 1 shows the location described above and the soil sampling points distributed in the study area.

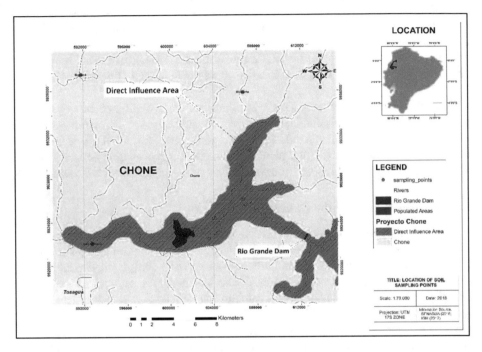

Fig. 1. Location of the study area and soil sampling points.

The canton Chone has a total area of 305 389.11 ha, 53.52% (163 441.70 ha) of this area is occupied with cultivated pastures, in the second order of importance depending on the surface can be found in the cultivation of cocoa with 13.26% (40 502.85 ha) of the total area of the Canton. Other important crops in this area are citrus fruits such as mandarin, orange, grapefruit, annual crops of corn and cassava, all of greater importance in the productive dynamics of the Canton (Instituto Espacial Ecuatoriano - IEE 2013).

Table 1 presents the optimal values for the different physical-chemical parameters of the relevant crops in the study area that were considered in the present work (Pantoja 2017).

2.2 Sampling and Soil Analysis

For the present work, soil sampling information was obtained from previous investigations in the study area, corresponding to Briones (2015) for the year 2013 and Calvopiña and Vilela (2017) for 2016. The samples obtained in 2016 correspond to 17 profiles, of which 9 were taken that coincide with the location of the profile samples obtained in 2013.

Table 1. Optimal values of physical-chemical parameters in crops

Farming Parameter	Banana	Cocoa	Corn	Grass
Ca^{2+} (cmol/kg)	3.0–9.0	1.5–4.0	3.0–6.0	2.0–6.0
CE (µS/cm)	300-600	200−400	300–500	200–500
Mg^{2+} (cmol/kg)	0.38–1.13	0.33–0.83	0.63–1.17	0.38–0.75
K^+(cmol/kg)	0.32–0.82	0.23–0.41	0.41–0.62	0.32–0.64
Na^+ (cmol/kg)	<0.61	<0.61	<0.61	0.04–0.07

Sampling was carried out using the zigzag method, which according to NOM-021-SEMARNAT-2000 [14], consists in taking the sample at the beginning of one side of the land, taking the starting point at random and thus defining a sampling homogeneous surface with a depth of 0.20 m, taking approximately 1.5–2 kg. The parameters analyzed by the accredited laboratory were: sodium (Na^+), magnesium (Mg^{+2}), calcium (Ca^{+2}), potassium (K^+), hydrogen potential (pH) and electrical conductivity (CE), with the methods presented in Table 2 for the year 2013 and in Table 3 for the year 2016.

Table 2. Physical-chemical determinations for soils (2013)

Parameter	Method
Calcium (Ca^{+2})	PEE-LASA-FQ-01-D/APHA 3500-Ca B
Electrical Conductivity (CE)	APHA 2510
Magnesium (Mg^{+2})	APHA 3500-Mg B
Potassium (K^+)	PEE-LASA-FQ-20a/APHA 3111-K B
Potential of Hydrogen (pH)	APHA 4500-H+
Sodium (Na^+)	PEE-LASA-FQ-20a/APHA 3111-Na B

Table 3. Physical-chemical determinations for soils (2016)

Parameter	Method
Calcium (Ca^{+2})	NOM-021-RACNAT-2000
Electrical Conductivity (CE)	NOM-AA-93-1984
Magnesium (Mg^{+2})	NOM-021-RACNAT-2000
Potassium (K^+)	NOM-021-RACNAT-2000
hydrogen potential (pH)	Direct electrode measurement
Sodium (Na^+)	NOM-021-RACNAT-2000

2.3   Statistical Analysis

With the analyzed data of the different physicochemical parameters that intervene in soil fertility, a correlation and linear regression analysis was performed to identify the degree of association of the variables. To verify results veracity, a hypothesis test was carried out with all the data obtained.

3   Analysis of Results

3.1   Linear Regression

Figure 2 shows the linear regression executed in software R for the years 2013 and 2016 in the nine sampling points for the potassium ion and Fig. 3 shows the linear regression executed in software R for the years 2013 and 2016 in the nine sampling points for the calcium ion. It is important to mention that the same analysis was carried out for magnesium and sodium ions.

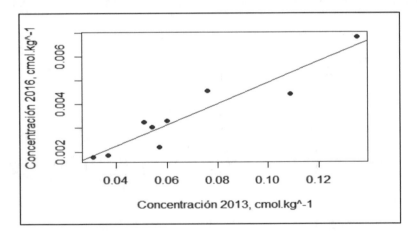

Fig. 2.   Linear regression of the concentrations in the potassium ion, years 2013 and 2016

In the linear regression obtained in Fig. 2, a decrease in nutrient concentration can be interpreted in 2016 with respect to the concentration in 2013. To verify the correct use of the linear regression method, a residue test can be observed that can be observe in Fig. 4, the result of this figure shows that there is no data distribution, therefore, the linear regression applied to the nine sampling points was successful. It is important to mention that the same residue test was applied for the linear regression of calcium, magnesium and sodium ions.

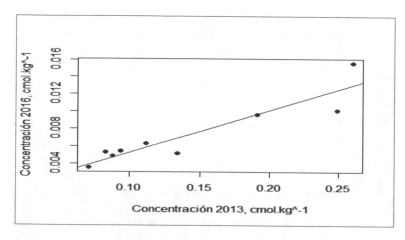

Fig. 3. Linear regression of the concentrations in the calcium ion, years 2013 and 2016

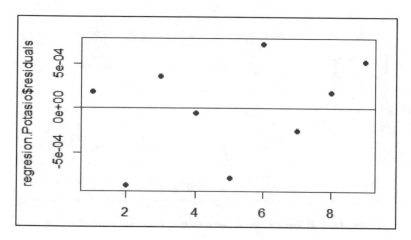

Fig. 4. Residue test for data correlation in potassium ion

Magnesium and Sodium ions, obtaining satisfactory results that, given that the correlation values are between 0.85 to 0.99, which indicates an almost perfect correlation according to Lind et al. (2012), who establish that a correlation coefficient of −1.00 or +1.00 indicates a perfect correlation.

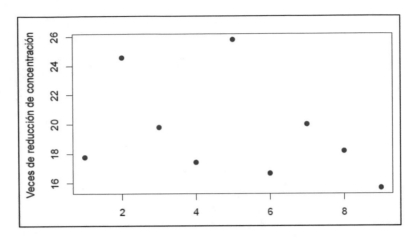

Fig. 5. Reduction of concentration by potassium ion hypothesis test

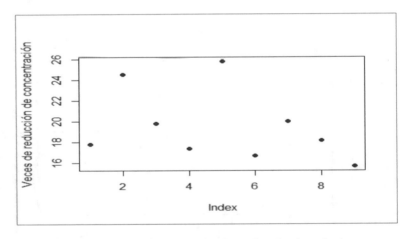

Fig. 6. Reduction of concentration by sodium ion hypothesis test

3.2   Statistical Analysis

Obtain a quantitative panorama of soil degradation in the years 2013 and 2016. A statistical analysis was carried out, which is presented in Table 4. The same analysis was carried out for calcium, magnesium and sodium ions.

Table 5 shows the average percentage of ions losses of Calcium, Magnesium, Sodium and Potassium in the years 2013 and 2016

3.3   Comparison of Measured and Required Nutrients for the Soil

Table 6 shows the resume of the results obtained by comparing the existing ions in 2016 with respect to the optimal amount of nutrients according to the main crops in the area.

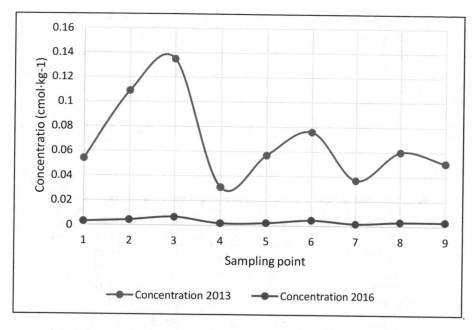

Fig. 7. Variation of potassium concentration in the years 2013 and 2016

Table 4. Deterioration in the concentration of the K ion in the years 2013 and 2016

Sampling point	Concentration 2013 (cmol·kg^{-1})	Concentration 2016 (cmol·kg^{-1})	Deterioration of concentration (%)
1	0.054	0.00304	94.37
2	0.109	0.00444	95.93
3	0.135	0.00682	94.95
4	0.031	0.00178	94.26
5	0.057	0.00221	96.12
6	0.076	0.00456	94.00
7	0.037	0.00185	95.00
8	0.060	0.00330	94.50
9	0.051	0.00325	93.63

Table 7 shows the deficit of nutrients existing in 2013 and the later deficit reached at 2016, these values are compared with the optimal values for the banana crop.

Nutrients in 2013 were deficient in relation to the amount required for banana cultivation, one of the most important ions in the banana crop the potassium ion only reached 11.89% of the requirements for an optimal banana crop. Despite the lack of nutrients in the soil in 2013, several crops could be found in the study area.

Table 5. Average deterioration of the ions in the years 2013 and 2016

Ion	Average concentration 2013 (cmol·kg^{-1})	Average concentration 2016 (cmol·kg$^{-1)}$	Deterioration of concentration (%)
K$^+$	0.06777778	0.00347222	94.88%
Ca^{2+}	0.14244444	0.00733111	94.85%
Mg^{2+}	0.06088889	0.00311778	94.88%
Na$^+$	0.135222222	0.006962667	94.85%

Table 6. Comparison of nutrients in 2016 compared to the optimum according to the crop

Ion	Average concentration 2016 (cmol·kg^{-1})	Optimal value for the crop(cmol·kg^{-1})			
		Banana	Cocoa	Corn	Grass
K$^+$	0.00347222	0.32–0.82	0.23–0.41	0.41–0.62	0.32–0.64
Ca^{2+}	0.00733111	3–9	1.5–4	3–6	2–6
Mg^{2+}	0.00311778	0.38–1.13	0.33–0.83	0.63–1.17	0.38–0.75
Na$^+$	0.006962667	<0.61	<0.61	<0.61	0.04–0.07

Table 7. Comparison of the ions needed in the soil for a banana crop, with the values measured in the years 2013 and 2016

Ion	Optimal values for Banana cultivation	Concentration measured in 2013	Percentage of the measured concentration over optimal values	Concentration measured in 2016	Percentage of the measured concentration over optimal values
Ca^{2+}	6.00	0.1424	2.37%	0.0073	0.12%
Mg^{2+}	0.76	0.0608	8.06%	0.0031	0.41%
K$^+$	0.57	0.0677	11.89%	0.0034	0.61%
Na$^+$	0.61	0.1352	22.17%	0.0069	1.14%

However, by 2016, the potassium ion requirements in relation to the optimum values for banana cultivation reached 0.61%. This situation Leaved the soil practically infertile and making it impossible to grow bananas in the study area.

Figure 8 shows the nutrient deficit for banana crop compared to the measurements made in 2013 and 2016.

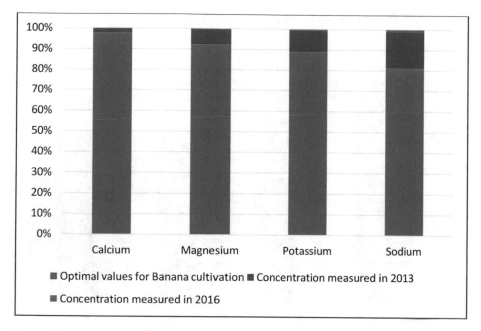

Fig. 8. Percentage of comparison of the ions needed in the soil for a banana crop, with the values measured in the years 2013 and 2016

Table 8 shows the deficit of nutrients existing in 2013 and the later deficit reached at 2016, these values are compared with the optimal values for the cacao crop.

Table 8. Comparison of the ions needed in the soil for a cacao crop, with the values measured in the years 2013 and 2016

Ion	Optimal values for Cacao cultivation	Concentration measured in 2013	Percentage of the measured concentration over optimal values	Concentration measured in 2016	Percentage of the measured concentration over optimal values
Ca^{2+}	2.75	0.1424	5.18%	0.0073	0.27%
Mg^{2+}	0.58	0.0608	10.50%	0.0031	0.54%
K^+	0.32	0.0677	21.18%	0.0034	1.09%
Na^+	0.61	0.1352	22.17%	0.0069	1.14%

Figure 9 shows the nutrient deficit for Cacao crop compared to the measurements made in 2013 and 2016.

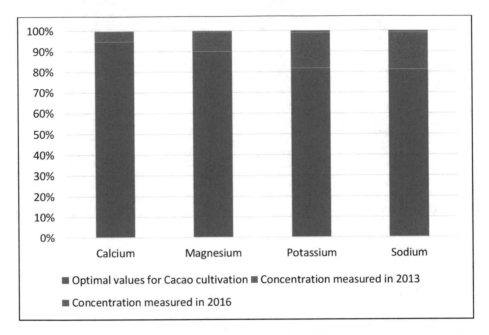

Fig. 9. Percentage of comparison of the ions needed in the soil for a Cacao crop, with the values measured in the years 2013 and 2016

Table 9 shows the deficit of nutrients existing in 2013 and the later deficit reached at 2016, these values are compared with the optimal values for the cacao crop.

Table 9. Comparison of the ions needed in the soil for a corn crop, with the values measured in the years 2013 and 2016

Ion	Optimal values for Corn cultivation	Concentration measured in 2013	Percentage of the measured concentration over optimal values	Concentration measured in 2016	Percentage of the measured concentration over optimal values
Ca^{2+}	4.50	0.1424	3.17%	0.0073	0.16%
Mg^{2+}	0.90	0.0608	6.77%	0.0031	0.35%
K^+	0.52	0.0677	13.16%	0.0034	0.67%
Na^+	0.61	0.1352	22.17%	0.0069	1.14%

Figure 10 shows the nutrient deficit for Corn crop compared to the measurements made in 2013 and 2016.

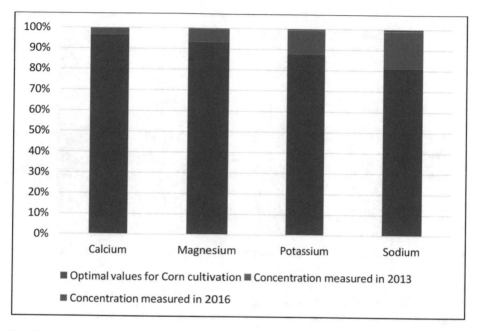

Fig. 10. Percentage of comparison of the ions needed in the soil for a Corn crop, with the values measured in the years 2013 and 2016

Table 10 show the deficit of the calcium, magnesium and potassium ions existing in 2013 and there later deficit reached at 2016, these values are compared with the optimal values for the grass crop. The value for the sodium was over the optimum values in the 2013.

Table 10. Comparison of the ions needed in the soil for a grass crop, with the values measured in the years 2013 and 2016

Ion	Optimal values for Grass cultivation	Concentration measured in 2013	Percentage of the measured concentration over optimal values	Concentration measured in 2016	Percentage of the measured concentration over optimal values
Ca^{2+}	4.00	0.1424	3.56%	0.0073	0.18%
Mg^{2+}	0.57	0.0608	10.78%	0.0031	0.55%
K^+	0.48	0.0677	14.12%	0.0034	0.72%
Na^+	0.06	0.1352	245.86%	0.0069	12.66%

Figure 11 shows the nutrient deficit for Grass crop compared to the measurements made in 2013 and 2016.

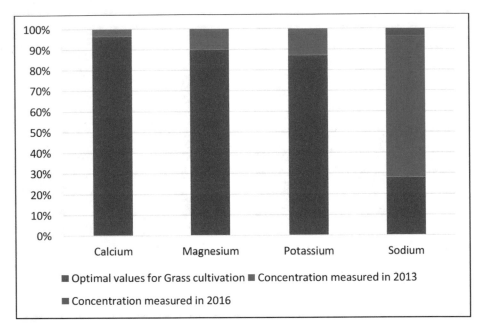

Fig. 11. Percentage of comparison of the ions needed in the soil for a Grass crop, with the values measured in the years 2013 and 2016

4   Conclusions

There is an almost perfect correlation (from 0.85 to 0.99) of the data, which indicates that the higher the concentration of nutrients, the greater the loss thereof, this was confirmed with the results obtained in the statistical analysis, which indicate a significant loss of nutrients. nutrients in the potassium ions (94.88%), calcium (94.85%), magnesium (94.88%) and sodium (94.85%) between 2013 and 2016, with the consequent loss of fertility in the soils of the study area, probably associated with the influence of the Río Grande reservoir of the PPMCH in the area, this phenomenon directly affects the agricultural production of bananas, cocoa, corn and grass crops. zone.

Although in the soils of the study area the loss of nutrients mentioned above was evident, these presented problems since 2013, because the optimal concentrations of nutrients for the crops grown in the area differ greatly magnitude with the values obtained from the concentration in the samples of the years 2013 and 2016, which indicates that the loss of nutrients may also be related to the agricultural activities developed in the area of direct influence of the PPMCH.

The parameters of pH (range of 6 to 8) and electrical conductivity (range of 350–650 μS/cm) measured in the samples of the soil profiles did not show significant changes between 2013 and 2016, maintaining optimum ranges for crops in area. For this reason, they were not considered in the statistical analysis performed.

By means of the hypothesis test it can be verified that the reduction of ions concentration between 2013 and 2018 turns into certain cases such as the potassium ion, up to 17 times, which is of great concern for an area in which its inhabitants depend on agriculture.

The decrease of nutrients in this area can generate problems of toxicity in soils and risks of health damage for people of the direct influence area, since soil acidification phenomena can occur due to changes in the concentrations of chemical species, which can increase the mobility of chemical pollutants, from the chemical substances used in the typical agricultural activities of Chone.

References

Andrades, M., Martínez, M.: Fertilidad del suelo y parámetros que la definen. Universidad de la Rioja, Servicio de Publicaciones, La Rioja, España (2014)

Arroyo, J.: La Ciencia Al Servicio De Las Políticas Ambientales En La Protección Del Suelo. Memorias del XXI Congreso Latinoamericano de la Ciencia del Suelo. Quito (2016)

Briones, M.: Determinación De La Salinidad Del Agua Y Suelos Del Proyecto Propósito Múltiple Chone, Manabí - Ecuador. Universidad de las Fuerzas Armadas, Departamento de Ciencias de la Tierra y la Construcción, Sangolquí, Ecuador (2015). http://repositorio.espe.edu.ec/xmlui/bitstream/handle/21000/12722/T-ESPE-049761.pdf?sequence=1&isAllowed=y. Accessed 25 June 2018

Calvopiña, K., Vilela, A.: Diseño De Tecnosoles Para La Retención De Fosfatos En El Agua, De La Presa Propósito Múltiple Chone (Ppmch), A Partir De Muestras De Suelo Del Cantón Chone, Manabí, Ecuador. Universidad de las Fuerzas Armadas, Departamento de Ciencias de la Tierra y la Construcción, Sangolquí, Ecuador (2017). http://repositorio.espe.edu.ec/bitstream/21000/13389/1/T-ESPE-057342.pdf. Accessed 12 June 2018

Cárdenas, R., Erazo, M.: Ecología: Impacto de la problemática ambiental actual sobre la salud y el ambiente. Ecoe Ediciones, Bogotá (2013). Accessed 10 Aug 2018

Carrera-Villacrés, D., Guevara-García, P., Gualichicomin-Juiña, G., Maya-Carrillo, A.: Processes controlling water chemistry and eutrophication in the basin of Río Grande, Chone, Ecuador. In: Pérez-Soto, F., Rocha-Quiroz, J., Salazar-Moreno, R., Sepúlveda-Jiménez, D. (eds.) Ciencias Químicas y Matemáticas Handbook T-I, vol. 1, pp. 25–36. ECORFAN-México, S. C., Texcoco de Mora, México (2015). http://hdl.handle.net/20.500.11799/41146. Accessed 10 June 2018

Empresa Pública del Agua: Propósito Múltiple Chone se inaugurará el 24 de noviembre. Samborondón, Ecuador: EPA – EP (2015). http://www.empresaagua.gob.ec/proposito-multiple-chone-se-inaugurara-el-24-de-noviembre/. Accessed 19 Apr 2018

FAO: Carta Mundial de los Suelos. Roma, Italia, June 2015. http://www.fao.org/3/b-i4965s.pdf. 11 Sept 2018

GAD Chone: Gobierno Autónomo Descentralizado de Chone, Junio 2014. http://www.chone.gob.ec/pdf/lotaip2/documentos/pdot.pdf. Accessed 23 June 2018

Garcia, G.: Fertilización En El Cultivo De Maíz Blanco Amiláceo. Cusco, Perú (2013). https://www.agrobanco.com.pe/data/uploads/ctecnica/022-g-mab.pdf. Accessed 12 Sept 2018

García, J., Puigdefábregas, J.: Efectos de la construcción de pequeñas presas en cauces anastomosados del pirineo central. Cuadernos de Investigación Geográfica, **11**(1–2), 91–102 (1985). https://publicaciones.unirioja.es/ojs/index.php/cig/article/view/946/841. Accessed 8 July 2018

García, Y., Ramírez, W., Sánchez, S.: Indicadores de la calidad de los suelos: una nueva manera. *Pastos y Forrajes,* **35**(2), 125–138 (2012). http://scielo.sld.cu/scielo.php?script=sci_arttext&pid=S0864-03942012000200001. Accessed 11 June 2018

Inamhi; Fao: Estudio Hidrológico de inundaciones en la cuenca alta del río Chone (subcuencas: Garrapata, Mosquito y Grande), Febrero, 2008. https://issuu.com/inamhi/docs/chone. Accessed 20 June 2018

Instituto Espacial Ecuatoriano – IEE: Instituto Espacial Ecuatoriano, December 2013. http://ideportal.iee.gob.ec/geodescargas/chone/mt_chone_sistemas_productivos.pdf. Accessed 22 June 2018

Lind, D., Marchal, W., Wathen, S.: Estadística Aplicada A Los Negocios Y La Economía. McGraw – Hill, México (2012). Accessed 24 June 2018

Pantoja, J.: Importancia De Conocer Los Límites Óptimos De Nutrientes En El Suelo Y Foliares Para La Nutrición De Cultivos. Quito, Ecuador: AGN Latam (2017). Accessed 15 June 2018

Secretaría de Medio Ambiente y Recursos Naturales: nom-021-semarnat-2000: norma oficial mexicana, que establece las especificaciones de fertilidad, salinidad clasificación de suelos. Estudios, muestreo y análisis. Diario Oficial de la Federación, México (2002). http://legismex.mty.itesm.mx/normas/rn/rn021-02.pdf. Accessed 22 June 2018

Vele, S.: Estudio De La Fertilización Del Cultivo De Cacao Nacional, En La Provincia De Santa Elena. La Maná, Cotopaxi, Ecuador, December, 2013. Accessed 25 Aug 2018

Yrízar, A., Búrquez, A., Calmus, T.: Disyuntivas: impactos ambientales asociados a la construcción de presas. Región y Sociedad, **24**(3) (2012). http://www.scielo.org.mx/pdf/regsoc/v24nspe3/v24nspe3a10.pdf. Accessed 4 July 2018

Methodology for Microgrid/Smart Farm Systems: Case of Study Applied to Indigenous Mapuche Communities

Carolina Vargas[1], Raúl Morales[1], Doris Sáez[1(✉)],
Roberto Hernández[1], Carlos Muñoz[2(✉)], Juan Huircán[2],
Enrique Espina[1], Claudio Alarcón[2], Victor Caquilpan[1],
Necul Painemal[3], Tomislav Roje[1], and Roberto Cárdenas[1]

[1] Universidad de Chile, Santiago, Chile
dsaez@ing.uchile.cl
[2] Universidad de la Frontera, Temuco, Chile
carlos.munoz@ufrontera.cl
[3] National Corporation for Indigenous Development, Temuco, Chile
npainemal@conadi.gov.cl

Abstract. In this work, a methodology for the design and implementation of a microgrid for supplying electricity to indigenous Mapuche communities in Chile is proposed. The challenge in designing this microgrid lies in making it compatible with a community's socio-cultural situation, thereby allowing active community participation in all stages of project development. To achieve this, we propose a participatory model of technological innovation that fits in with the local culture. The proposal is based on taking advantage of the communication and energy infrastructure of the microgrid to build a smart farm system to improve the quality of life of Mapuche rural communities. The smart farm consists of a set of technological devices and software applications related to communitarian farm activities that attempt to reduce the gap of information technologies. To demonstrate the potential of the smart farm system, we propose solutions for livestock monitoring and rational water use at the local level since these two applications represent needs expressed by the communities themselves. To develop a harmonious project consistent with the socio-cultural situation of the Mapuche people and to obtain effective community participation in the different stages of the project, we suggest performing a participatory socio-environmental diagnosis to identify the project's potential and existing initial barriers to its implementation in these communities. Subsequently, we will evaluate its potential impact on the social welfare and development of the communities. This design will result in a participatory project with community management for ensuring its sustainability under a concept of development shared by the community.

Keywords: Microgrids · Smart farm · Participatory model

© Springer Nature Switzerland AG 2019
J. C. Corrales et al. (Eds.): AACC 2018, AISC 893, pp. 89–105, 2019.
https://doi.org/10.1007/978-3-030-04447-3_6

1   Introduction

Access to electricity in rural zones has a strong impact on local development [1], resulting in direct benefits to the territory such as reducing poverty, stemming migration to the cities, promoting productive development and improving access to basic services such as health and education [3]. However, despite the fact that governments in Latin America have helped to improve the electricity supply to isolated areas with national and regional electrification programs the implemented projects have not considered the particular features of the local situations of isolated rural communities, including the dispersion of families, their socio-economic conditions, the lack of accessibility and connectivity and their cultural particularities [1–5]. Currently, most new electricity distribution projects consist of extending the conventional grid into rural areas. However, this solution involves high costs and technical difficulties [2, 6].

In this context, the use of Non-Conventional Renewable Energy (NCRE) as the principal power source is an attractive solution for isolated rural areas [7, 8]. This is because the costs in the operating stage are low compared to those of traditional electricity generation projects. Furthermore, NCRE produces a lower environmental impact in the area and contributes to the sustainability of the territory [3, 9]. These technologies, together with other types of small-scale generation, loads and energy storage systems operating as a single, controllable low-voltage system, form a microgrid, which provides a more stable and secure electricity supply. Internationally, there is experience with microgrid installations in urban sectors, e.g., in Denmark, the Netherlands, the United States, Japan, Canada and Spain [10], as well as in rural areas in Ecuador, South Africa, Morocco and Bangladesh. Microgrids also exist on islands in Greece, Ecuador and Mexico [10].

In the rural context, electrification projects with NCRE seek to increase the use of these natural resources and improve the quality of life of poorer populations. In recent years, there have been some developments of microgrids in Chile. Among them, microgrids based on Wind-Diesel generation in rural zones in Chile have been installed by Wireless Energy Company, for instance, in Tac Island, Chiloé, and Desertores Island [11]. Additionally, one microgrid based on renewable energy was installed in Huatacondo- Chile, Tarapacá Region [4]. This initiative within Latin America incorporated the active participation of the community, which is a fundamental aspect of the sustainability of the system [8]. This enabled the community to manage the system through participation in the operation and maintenance stages of energy management, including a co-management process until full self-management could be achieved [4, 5]. This approach also recognizes the sovereignty of communities over the natural resources in their surroundings [9]. This is important when implementing these systems in isolated rural communities with an ethnic identity since, unlike typical rural communities, the way of life and the world-view of indigenous peoples are directed by fundamental principles: that mankind is a part of the natural environment, that the welfare of the community predominates over that of the individual, and that strong spiritual links are expressed with nature, established as their world-view.

The Mapuche people is the largest ethnic group among the indigenous peoples of Chile (86.4%). The people identifies strongly with the above-described world-view.

The expectation therefore is to develop a technological project to promote a system that can be adapted to these socio-cultural characteristics for recovering essential components of these communities that are in danger of being lost [4]. It is also hoped that these communities will be willing to adopt this type of technology, mainly since one of the greatest benefits of access to electricity and the productive activities proposed under the smart farm will be to reduce migration from the community zones to the cities [5]. Simultaneously, for the microgrid implementation in southern Chile, it is important to find a community that will assimilate all stages of an electricity system project and participate in its construction. Thus, the expected final result is to implement the design in a community that will be a positive reference for other communities interested in incorporating this type of technology into their way of life. It is therefore of fundamental importance to distinguish the considerations or aspects to be considered when choosing a community in which to introduce this technology [5]. To this end, we will utilize the work of Energy Centre of Universidad de Chile on the identification of Chilean communities cut off from the national grid using technical, socio-environmental and political criteria [5]. This work presents a participatory-model-based design of a microgrid for electricity supply for an isolated Mapuche community, including the management and rationing of irrigation water and a livestock monitoring system under the smart farm concept. The proposal addresses both social and technical aspects to generate an integrated technological solution for improving the quality of life of Mapuche rural communities while strengthening the appreciation of their culture and ethnic identity. From a critical angle, there is no clear delimitation between what constitutes the impacts of microgrid/smart farm projects in the communities where they are implemented and the evaluation of the sustainability of the projects. The participatory model for technological innovation proposed in this project is intended to overcome these methodological deficiencies, its central axes being community participation and self-management, under the framework of gradual technological and social change, understood as an integrated system in harmony with the Mapuche community.

2 Background

2.1 Participatory Model for Technological Innovation

As the authors in [12] state, government officials and citizens have different worries and claims; officials stick to objective and factual claims, whereas citizens are more inclined to operate with social-world claims such as which energy source contributes more to the community. To avoid these differences, for the design and implementation of the microgrid and the smart farm system, we propose a participatory innovation model for local development consistent with the indigenous culture, which will ensure its sustainability. A project is considered to be sustainable when it achieves self-management in social, economic and environmental terms. It must therefore be implemented at the community level through open, participatory processes in which the opinions and expectations of the various actors are incorporated [4, 13].

The literature on the social aspects involved in microgrid projects worldwide focuses on methods for the economic and socio-cultural evaluation of these initiatives and their impacts on communities for sustainable local development. Sustainability is a key element that must be considered to ensure that projects are viable over time under the direction of the communities where they are implemented [5].

Analysis of the evaluations and impact variables used in various projects implemented in different locations around the world [7, 14] leads to the conclusion that the utilized methods have produced diverse results: services associated with electric power, the generation of new jobs, the development of tourism, education, health standards, drinking water, personal and community security, effects produced by television and changes in social relationships. This is an indicator of the lack of methodological evaluation consistency for an integrated focus on the economic, social, cultural and environmental dimensions of the studied communities, which needs to be developed.

From a critical angle, there is no clear delimitation between what constitutes the impacts of microgrid/smart farm projects in the communities where they are implemented and the evaluation of the sustainability of the projects. The participatory model for technological innovation proposed in this project is intended to overcome these methodological deficiencies, its central axes being community participation and self-management, under the framework of gradual technological and social change, understood as an integrated system in harmony with the Mapuche community.

2.2   Microgrids

Microgrids are considered to be a particular case of the general concept of a Smart Grid, an interdisciplinary term for a set of technological solutions for a power management system [15]. Distributed generation, in the form of small power sources installed close to consumption points, has become more commonly used because of its low contaminant emissions and operating costs..

Microgrids are a set of loads and small generators operating as a single controllable system that provides power and/or heating to the associated local area. The energy sources used in grids of this type are small generators (<100 [kW]) with electronic interfaces such as micro-turbines, photovoltaic panels and fuel cell. Microgrids can have two operating modes: connected to the main grid and isolated from it [16]. For secure operation, a microgrid must have two or more sources of distributed generation, and an Energy Management System (EMS) is required to operate in isolated mode. For control purposes, the connection/disconnection state of the microgrid can also be included. When the microgrid is operating isolated from the main grid, its dynamic depends on the micro-generators, the control system and the grid. Local generation follows the load, and storage and load-sharing units are normally used to increase grid security [16]. The most important control requirements were proposed, i.e., using local micro-generator controllers, optimizing microgrid operation while minimizing costs, and utilizing protection systems in the distribution grid.

The development of microgrids to satisfy local power needs has expanded considerably in recent years. The innovative feature of the present work is that, apart from the technical aspects usually considered, it produces a design that is in harmony with the indigenous culture of the selected area, therein considering aspects of cultural identity to obtain a sustainable microgrid project, with much of the project's value added by the Mapuche community. The involvement of the community is essential, as [17] states; an innovative environment facilitates the success of the initiative but requires a strong commitment by the actors linked to the project. In light of these considerations and taking advantage of the communication and energy infrastructure of the microgrid, we propose to implement a smart farm system for the management of the available water resources and a livestock monitoring system.

2.3   Smart Farm System

The set of communication and power generation technologies involved in the construction of a microgrid has enabled us to introduce the smart farm concept (SMART = Sustainable, Manageable and Accessible Rural Technologies [18]). In this project, the smart farm is intended to monitor different areas of importance for agriculture using soil humidity measurements, irrigation water flow sensors, well level sensors, and livestock location devices to both improve the performance and use of natural resources and promote productive development. These initiatives are oriented towards agricultural systems established for monocultures [19], but other systems have been proposed for domestic use, therein emphasizing the use of renewable energy [20]. The smart farm concept can be oriented towards isolated rural communities with models of production operating in harmony with the environment [21], thereby making it possible to use a social participation model to design the tools with an inclusive cultural approach [22].

   Cattle rustling has a serious impact on the life of Mapuche families, and therefore, cattle supervision is important to these communities. Various monitoring systems collect information on the location of stock in real time, therein being specially designed for purposes such as stock control and virtual fences, virtual fences for grazing management with GPS and WiFi or RFID devices and ZigBee-type wireless networks for locating animals on grazing land [23–31]. Nevertheless, these systems do not completely solve the problem proposed here because they do not incorporate cultural aspects such as restrictions on the installation of devices (ear tags, collars, and antennas); cost limitations, considering that indigenous stock-rearing is a subsistence activity; and operational aspects limiting the amount of maintenance that can be conducted in the field. On the other hand, Mapuche families do not need to know exactly where every animal is, only in which general area, to coordinate with their neighbors to facilitate community action. These solutions need to be socially and economically sustainable. In [25], the sustainability and applicability of monitoring systems that take into account the previous considerations were analyzed, therein showing that the compromise between economic and social sustainability is important but that a participatory approach is needed to analyze in depth the multiple relations (compromises and synergies) generated to evaluate projects of this type.

   Another important application of the smart farm, proposed for Mapuche communities, is related to the rational use of irrigation water, which is important for the subsistence of isolated rural communities [22]. There is a need to regulate water extraction, therein defining clear systems for monitoring water extraction, models for rational water use and participatory regulation systems. Although there are models for studying the availability of subterranean waters [26], they are not general models, and specific cases need to be studied [27] using information captured in the field at regular time intervals. The fact that the design of the microgrid/smart farm system includes cultural aspects from its inception, and that its approach is community based rather than individual based, differentiates this proposal from other proposals, therein letting the community as a whole take actions under certain critical conditions such as the loss of livestock and the rational use of water.

3   Design of a Participatory-Model/Microgrid/Smart Farm System

The design of the participatory-model/microgrid/smart farm integrated system includes stages to consider for developing the proposal, as will be explained in the following sections. It should be noted that each of the considerations proposed in each stage will be adapted for functioning within a Mapuche rural community.

3.1   Integrated System

Figure 1 shows the outline of the integrated system to be applied in the Mapuche communities, where the microgrid provides an innovative way of incorporating a smart farm system that includes water use management and livestock monitoring to produce a positive impact on the use of Mapuche lands. In the first instance, the proposal involves selecting the most suitable community for adopting this system using social and technical criteria adapted for analysis in the Mapuche context. Once the selection process is completed, a participatory analysis must be conducted through the collection of data on the Mapuche community selected to evaluate cultural, economic, social, environmental and technical aspects towards identifying the existing potentials and constraints that will define the system.

The power supply distributed through the microgrid and its communications platform will be used jointly with systems for monitoring livestock and efficient water management, with the aim of consolidating the community's productive activities. Each stage of this project must be designed based on the application of field techniques and methods in the framework of a participatory model for technological innovation to actively involve the community in all processes.

3.2   Participatory Model for Technological Innovation

To ensure the social sustainability of the project, in the context of management shared with the rural community, we propose to develop a social participatory model, shown in Fig. 2, with the following stages:

1. Technical-social criteria for selecting a community: To adopt this system, the most appropriate communities must be identified.

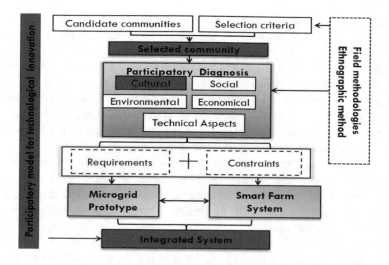

Fig. 1. Stages and processes of the integrated system proposal.

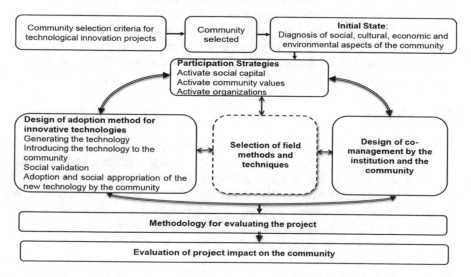

Fig. 2. Diagram of the proposed participatory model of technological innovation for local development.

2. Initial state: The initial state of the community for implementation must be determined, including cultural, social, economic, socio-environmental and technical aspects.
3. Participation strategies: To obtain effective participation by the selected community in all stages of the technological innovation project.

4. Selection of field methods and techniques: The Ethnographic Method, a basic tool of Anthropology, will be used to determine the appropriate methods and techniques for field work.
5. Design of a method for the adoption of innovative technologies: This process of technological change implies first developing the new technology in the laboratory and then introducing it to the community. Next comes the social validation, which occurs when the community sees, analyses and evaluates the attractiveness of the innovation. Finally, decision making and the incorporation of the innovation into the socio-cultural system can be achieved.
6. Design of co-management: For this project, co-management by the institution or organization as an agent for technological change and the community where it is implemented is of primordial importance because this will allow the community to develop the capacity for self-management through training, without which the project will be unsustainable.
7. Methodology for social evaluation of the project: Follow-up and evaluation of a project are necessary to identify the weaknesses in functioning and enable the reorientations and adjustments necessary to achieve the proposed objectives. The participation of the community where the project is implemented, as part-responsible, is indispensable to ensure the sustainability of the initiative.
8. Determining the impacts of the project on the community: A methodology with appropriate indicators must be developed to determine the impacts of the project on the socio-cultural, economic and socio-environmental aspects of the community.

4 Case Study

The proposed methodology was applied to the Mapuche community of Huanaco Huenchun in the district of Nueva Imperial, Araucanía Region (Chile). The community consists of two sectors, Imperialito and Puente Fierro, with 83 families and 185 inhabitants. Semi-structured interviews, surveys of each family, observations and participatory workshops were conducted to obtain information on social, cultural, economic, environmental and technical aspects, thereby forming the initial diagnosis of the community and characterization of the selection criteria. Participatory diagnosis was used to develop the design of the microgrid, therein incorporating the smart farm system, which was intended to be consistent with the Mapuche culture.

4.1 Technical-Social Criteria for Selecting a Mapuche Community

A methodology is developed for inclusion in projects of this nature in which the first stage must be the identification of the most suitable communities for the adoption of the proposed technologies. This is because the communities must participate in all stages of the project, from design to operation, maintenance and evaluation. From the point of view of financing, many rural electrification schemes have limited resources; therefore, it is necessary to give priority to communities where the projects can actually be implemented. This prioritization implies establishing a series of aspects or criteria with respect to the sustainability of the initiative in the medium and long term, which are generally linked to the objective of each specific project. However, there are certain common characteristics that are required in a community for the introduction of a technological project in which it is expected that the community will appropriate the system. Below is a list of the aspects that should be considered when selecting a community for implementation of the participatory-model/microgrid/smart farm system [5].

Poverty Level. The aim of the project is to contribute to satisfying the needs of a Mapuche rural community. This means providing a technological system that will have a positive impact on local development, with efforts focused on human groups living in poverty. The Mapuche people have a concept associated with poverty called "fillan", which is used to describe a condition of shortages of, for example, food from their own production unit or when people have no way of heating their homes (shortage of firewood). In general, most of those interviewed claimed to be content with their current situation. However, the shortage of land (average of 2.68 ha per family) and the low availability of irrigation water placed limitations on development.

Community Participation. For any type of project to be successful according to project developers, it is important for the community to take an active role in implementing the various stages of the proposal [6, 13]. The organizations present in Mapuche communities are divided between functional organizations, linked to the Chilean government, and traditional bodies. In the traditional bodies, it is important to recognize authorities, such as the Lonko (chief), the Machi (spiritual leader), the Werken (councillor/spokesman) and the Kimche (sage), since they are the ones who manifest and practice the Mapuche world-view and lead and thus can organize the community. This is fundamental for a project that relies on active community participation and building on the Mapuche culture. The results for this criterion showed that 90% of those interviewed belong to community organizations. All interviewees stated that they participated in meetings and discussions in these organizations, indicating a high degree of community participation. The only traditional authority recognized in these organizations was the Lonko as leader for cultural matters. The roles of committee representatives in the community were also found to be important for community development.

Social Cohesion. Social cohesion is considered one of the most important criteria for a project to achieve its objectives and be sustainable [28] since it reflects the community's internal functioning, related to the recognition, validation and respect of individuals for their peers and of the community's identity. Among the Mapuche people, social cohesion is defined through its cultural expressions, such as the existence of ceremonies and rites in the community. The results indicate good relationships between community members and local institutions, mainly the Municipality. In terms of aspects connected with the Mapuche culture, it was found that important ceremonies, continue to be held by the community, indicating that there remains a sense of Mapuche identity and that the people share the Mapuche world-view, a positive factor for the success of the project.

Demographic Dimension. The self-management of a technological project involves various functions and roles that must be determined jointly with the community so that they are accepted and validated by all its members. The limitations of these tasks must be borne in mind so that they can be assigned appropriately [5]. Certain aspects connected with the community's view of gender, population mobility and age-related roles therefore need to be considered in decision making. The above aspect is particularly important in Mapuche communities because the family, with the roles and functions characterizing family members, defines the structure of production and consumption among the Mapuche people [29]. The demographic make-up of the community was described by age, gender and mobility. There is a small percentage of young people living in the community (21%), which reflects the generalized migration occurring in the Chilean countryside. The percentage of adults in the community (44%) determines the existence of a potentially important human resource for the implementation of a microgrid in the community. Fewer than 30% of community members move away. The gender ratio is relatively homogeneous, with 52% female and 48% male. In terms of gender roles, women spend most of their time at home as housewives. Some women cultivate vegetable gardens or rear animals, such as domestic fowl, and may sell their production in the nearest major town. The head of the household in the surveyed families is the man, who works mainly in agriculture labors.

Power Generation Potential. The communities where the estimated energy potential from renewable resources will satisfy the estimated electricity demand of the community must be identified. This criterion applies to rural indigenous communities; however, it is important to analyze whether a resource that might potentially be used for energy could affect the tutelary or protecting spirits of the Mapuche. Specifically, it is advisable to investigate the relationship of the community with water bodies (mini-hydro schemes) and the forest (biomass), which may be the home of the newen or tutelary spirits of water and natural medicines.

First, the selection criteria described in Sect. 3 were applied to rural communities of the Araucanía Region. The community of Huanaco Huenchun proved to be the most suitable for ensuring a successful project. This decision was based on the following:

4.2   Participatory Diagnosis

The participatory diagnosis or initial state of a community is a baseline from which to analyze all the community's different dimensions – socio-cultural, economic, socio-environmental and technical – which are fundamental for ensuring that the design is in accordance with the local situation. The results of the selection criteria were used to make a diagnosis, conceived as the "initial state" of the community.

Cultural Ambit. The elements of Mapuche culture alive in the community are conceived as forms of adaptation to the conditions of its natural and social environment. This ambit includes their conception of the natural surroundings and the value placed on them, the ways in which social relations and leadership are organized, concepts and values attached to economic-productive activities, their world-view and religious beliefs. A review of these cultural aspects demonstrated that a local expression of

Mapuche culture existed but was weakened by the loss of important elements such as the use of the traditional language, Mapudungun.

Social Ambit. For this aspect, it is important to make a "social x-ray" of the community, which will include recognition of the social organizations present, the degree of community participation, social cohesion, education level and demographic aspects such as gender, age break-down and population mobility. In addition to the results of the criteria referring to participation, social cohesion and demographic aspects, it should be noted that 47% of the inhabitants between 18 and 65 years of age had completed secondary education. This information is important for analyzing and proposing suitable methodologies in the training workshops, which form part of the co-management stage of the project in the context of a participatory model. The main occupation of the inhabitants of the community is peasant-type family farming directly associated with arable crop production and to a lesser degree livestock-rearing.

Economic Ambit. Peasant-type family farming is performed in the community. Work is performed principally by members of the family, who are both the producers and consumers of the typical crops of interior dry-lands. Soft fruits and vegetables are produced in vegetable gardens near the houses. Most of the families sell part of their agriculture production, which represents the main cash income for the majority of the households surveyed. Families that raise livestock use it for their own consumption and sometimes for sale. In summer, the wetland is used as a community grazing area since it is fertile pasture. One of the main problems with livestock-rearing is rustling, especially during the summer months, as the pasture area is located a large distance from the owners' dwellings. In winter, people fish freshwater crabs for their own consumption and for sale. Another serious problem for the community is water shortages in summer since drought affects all these economic activities. The inhabitants therefore draw water from wells near their houses. In summer, however, the wells are unable to supply the normal needs for irrigation and water for the animals.

Environmental Ambit. The community of Huanaco Huenchun covers an area of 435.99 ha and is located around a wetland or marsh, which in winter is a seasonal habitat for endemic wild birds such as black-necked swans and some species of heron. The birds are subject to illegal hunting, as there is little environmental protection of the wetland. Neighbors are also affected by noise, damage and the risks involved in shooting. The wetland is a flat, low-lying, bare prairie that covers an area of 58.9 ha and it is waterlogged for much of the year. This has resulted in some productive activities, such as potato-cropping in the semi-humid ground surrounding the wetland and dry-land cultivation (wheat, beans, and peas), in sectors further away from the river.

Technical Aspects. The community is split between the administrative districts of Nueva Imperial and Carahue. These towns are connected by road S-40, which crosses the community and provides access by interurban public transport. The distance between dwellings varies between 20 and 845 m. Almost all the families are connected to the electricity main grid, which runs through much of the community, as shown in Fig. 4. However, there are frequent power outages, and the service is expensive. The dwellings are concentrated along the roads, as these provide transport for people and goods to the neighboring towns.

4.3 Design of Microgrid/Smart Farm System

The microgrid power supply system will be integrated with a smart farm system to promote the development of sustainable, controllable, and accessible technologies in rural environments [19]. The applications based on the needs proposed by the community have two components: on-line livestock monitoring and the remote measurement of the water levels in wells drawing on surface aquifers (rational water use).

To install the system, a wireless communications network that covers all sectors of the community must be set up [30]. To this end, the community will be connected with a set of ZigBee wireless routers linked to a coordinator that downloads all the data from the small-scale server and allows information to be uploaded to the internet. This device must incorporate a mini web-server providing messaging with important information concerning the Microgrid, the wells and livestock monitoring systems.

The solution proposed to support Mapuche families facing the problem of livestock rustling or theft is designed for very small communities with subsistence stock-rearing; therefore, both the cost and maintenance must be minimized. The device must be designed as a capsule that will be attached to livestock, for example, by incorporating the device into the collars worn by cows, horses or sheep [24]. In the proposed design, the owner wants to know whether his livestock is in the community area and its whereabouts. This system must be very low maintenance; therefore, autonomous power generation systems need to be considered. For rational water use, we suggest generating a system for the remote measurement of the power consumed by the pumps and ultrasonic or equivalent sensors to measure the water level in the wells. In addition to these instruments, models will be generated for determining whether there is any correlation between the water levels in families' wells and the weather forecasts. These water consumption models, together with the community's own water use policies, will allow the use of electric pumps to raise water. Efficient use of power for pumps is also important for managing the microgrid considering they are programmable loads. The Mapuche community selected will determine, among other things, the availability of the natural resources to be used as the power generation source, therein allowing the power potential to be identified as consistent with the available technologies. Using the participatory model, aspects that define local needs and constraints, such as the number of families, their principal economic activities and the geographical distribution of the dwellings (dispersion), will also be identified. These considerations define the profile of the electricity demand for domestic use and common areas, which will determine the requirements for the microgrid. When this information is matched with the power potential, the feasibility and viability of the project can be evaluated under technical, economic, social and cultural criteria. Possible microgrid solutions will be designed considering three possible architectures: centralized, decentralized and distributed.

In the design of a centralized microgrid, a single microgrid is required to supply all the dwellings in a defined zone (low geographical dispersion). Microgrids with a decentralized architecture provide local solutions, for example, photovoltaic panels and energy storage for each dwelling. This solution would be suitable for locations where the dwellings are widely dispersed and where a centralized solution would be more costly and less efficient. The size and topology of the microgrid were designed using the participatory diagnosis and considering the data on both the community's consumption to be supplied and the power potential available at the site.

Due to the difficulty in obtaining real consumption data, a methodology was implemented to determine the community's domestic demand by combining sample readings from smart meters with the socio-economic information on the residents obtained in semi-structured interviews. This methodology is based on identifying demand patterns that associate the socio-economic characteristics of residents with some demand profiles. The installed smart meters make it possible to obtain consumption data with a high time resolution (5 min in this case), thus allowing details of variations in demand within a particular period to be recorded. The surveys were drafted to obtain important data reflecting the behavior of members of the family group, which in turn affects demand. Data were obtained on the family composition of the households, age and type of activity performed by household members and the type and use of electrical appliances. A non-supervised classifier called Self-Organizing Maps (SOM) [31] was used to identify common patterns or classes. This tool was selected due previous expertise and for its great versatility in representing output elements. SOM allowed key aspects to be identified, such as the types of loads present in the community and the periods of both maximum demand and increased demand in the long and medium terms.

Although a river borders the community, its characteristics and the existing water rights mean that it is not feasible to install mini-hydro plants. To assess the potential for wind and solar power in the area, wind speed and horizontal solar irradiance in the community were measured over a period of 5 months. The community receives its electricity supply from a distribution utility; however, there are frequent power outages, and thus, the microgrid must be able to maintain the power supply when faults occur. The existing distribution grid will be used to satisfy consumption with the generated power. For sizing the microgrid system, the HOMER-PRO Energy software was used, in which the consumption and power resource profiles obtained previously are entered to obtain the optimum configuration according to the models of the chosen wind generators (result: 3 kW), photovoltaic panels (result: 18 kW) and battery banks. As shown Table 1, the investment cost for this configuration is US$51,726 since there is no need to install a distribution grid. Photovoltaic generation is the most costly type, accounting for almost 55% of the total investment, as shown in Table 1. The most expensive element of operation and maintenance (O&M considering 20 years for evaluation) is the grid since the community needs to purchase electricity at a total cost of US$99,326. Under this configuration, the power cost is US$0.178 per unit, lower than the cost paid by the community before the microgrid (US$0.246).

To select the best locations for the generation equipment connected to the microgrid, technical-social criteria were applied, and three sites in the community were selected as a preliminary recommendation. The community was consulted to determine the final locations, therein involving community members in the design of the microgrid and avoiding conflicts with their productive activities, ceremonies and sacred places. For the planning and monitoring of the microgrid operation, smart meters (of the same type used to measure the demand) that are able to transmit measurement signals to the microgrid main controller or EMS through the same communication network used by the smart farm system were considered.

Table 1. Costs (USD) of the microgrid for the case study in the community of Huanaco Huenchun, Chile.

	Investment	O&M	Total
Photovoltaic panels	28,338	3,104	31,442
Wind generator	14,185	2,483	16,668
Converter	9,203	1,007	10,210
Distribution grid	–	99,326	99,326
Total	51,726	105,920	157,646

For the smart farm system, a main wireless communication line (backbone) was designed for installation in the community. Where possible, communications routers were attached to power line posts and fence posts. In the wetland area, the communications system was designed using antennas specially adapted for use in flood zones.

To address the rustling problem, wireless tracking devices were designed to be attached to the livestock. These devices will communicate with the antennas of the communications backbone. This system will allow the owner to see whether an animal is inside the community area, and in which sector, by observing which antenna receives the strongest wireless signal from the animal's transmitter. From the study of the community geography, we observed that the network communications have to cover the summer and winter grazing areas as well as the access routes into the community from the road system. To monitor the water level in wells, as part of the smart system, we propose an ultrasonic device that also measures the activity of the well pump. To identify the times when domestic power consumption is drawing energy generated from renewable resources, the design includes the installation of indicator devices in the houses and wells, communicated wirelessly to the EMS. They will indicate when generation allows for high, medium or low power consumption and provide information about animals, well levels, and data from meteorological sensors.

We propose the use of ZigBee modules mounted on micro-controllers for the implementation of devices such as collar tags, well-monitoring devices and the trunk-line communications antennas of the microgrid/smart farm system. The cost of the wireless communications backbone designed to meet the requirements of the microgrid is US$3,675. In terms of the number of animals, the hardware used allows a maximum of 256 wireless devices per antenna, thereby allowing ample margin for new monitoring devices to be introduced. The total costs for the smart farm system (US$11,100) are less than 10% of the costs of the microgrid.

5 Conclusions

This work presents an innovative design for a microgrid/smart farm system, with a participatory model of technological innovation using a microgrid to supply the electricity demands of the community and the power and communications system required by the smart farm system. This integrated system will improve the quality of life of these Mapuche communities through rational water use and by monitoring their livestock in real time, thus responding to the needs of the communities. Additionally, it considers cultural aspects in the design of the proposal and the selection of energy resources to be used for generation. A fundamental element of this proposal is the design of criteria for selecting isolated rural communities in which to apply the system, therein considering both the situation of these indigenous Mapuche communities and the participatory model applied throughout all stages of the project. This selection method will reduce the uncertainty in the project operation and sustainability, as the selected community satisfies the established criteria. A case study of the design of a participatory-model/microgrid/smart farm System for the Huanaco Huenchun Mapuche community is presented as an example of this methodology. The following step is to implement the microgrid/smart farm, which is been addressed in a new project.

Acknowledgements. This work has been partially supported by FONDEF ID14I10063 project "Design and Implementation of an Experimental Prototype of Microgrid for Mapuche Communities". Also, this paper has been partially supported by Complex Engineering Systems Institute (CONICYT – PIA – FB0816).

References

1. Niez, A.: Comparative study on rural electrification policies in emerging economies: key to successful policies. Inf. Pap. p. 118 (2010)
2. Pereira, M.G., Freitas, M.A.V., da Silva, N.F.: Rural electrification and energy poverty: Empirical evidences from Brazil. Renew. Sustain. Energy Rev. **14**(4), 1229–1240 (2010)
3. Ferrer-Martí, L., Garwood, A., Chiroque, J., Ramirez, B., Marcelo, O., Garfí, M., Velo, E.: Evaluating and comparing three community small-scale wind electrification projects. Renew. Sustain. Energy Rev. **16**(7), 5379–5390 (2012)
4. Alvial-Palavicino, C., Garrido-Echeverría, N., Jiménez-Estévez, G., Reyes, L., Palma-Behnke, R.: A methodology for community engagement in the introduction of renewable based smart microgrid. Energy. Sustain. Dev. **15**(3), 314–323 (2011)
5. Ubilla, K., Jimenez-Estevez, G.A., Hernadez, R., Reyes-Chamorro, L., Hernandez Irigoyen, C., Severino, B., Palma-Behnke, R.: Smart microgrids as a solution for rural electrification: ensuring long-term sustainability through cadastre and business models. IEEE Trans. Sustain. Energy **5**(4), 1310–1318 (2014)
6. Baldwin, E., Brass, J.N., Carley, S., MacLean, L.M.: Electrification and rural development: issues of scale in distributed generation. Wiley Interdiscip. Rev. Energy Environ. **4**(2), 196–211 (2015)
7. Camblong, H., Sarr, J., Niang, A.T., Curea, O., Alzola, J.A., Sylla, E.H., Santos, M.: Microgrids project, Part 1: Analysis of rural electrification with high content of renewable energy sources in Senegal. Renew. Energy **34**(10), 2141–2150 (2009)
8. Palma-Behnke, R., Ortiz, D., Reyes, L., Jimenez-Estevez, G., Garrido, N.: A social SCADA approach for a renewable based Microgrid - The Huatacondo project. In: 2011 IEEE Power and Energy Society General Meeting, pp. 1–7 (2011)
9. Leary, J., While, A., Howell, R.: Locally manufactured wind power technology for sustainable rural electrification. Energy Policy **43**, 173–183 (2012)
10. Huayllas, T.E.D.C., Ramos, D.S., Vasquez-Arnez, R.: Microgrid systems: current status and challenges. In: 2010 IEEE/PES Transmission and Distribution Conference and Exposition: Latin America, vol. 900, pp. 7–12 (2010)

11. Stevens, N., Bergey, M.: Wind/Diesel microgrid utility service in desertores islands. In: Microgrid Deployment Workshop (2013)
12. Fast, S.: A Habermasian analysis of local renewable energy deliberations. J. Rural Stud. **30**, 86–98 (2013)
13. Walker, G., Cass, N.: Carbon reduction, 'the public' and renewable energy: engaging with socio-technical configurations. Area **39**(4), 458–469 (2007)
14. United Nations Development Programme, Towards an "Energy Plus" Approach for the Poor: An Agenda for Action for Asia and the Pacific, Bangkok, Thailand (2012)
15. Olivares, D.E., Mehrizi-Sani, A., Etemadi, A.H., Cañizares, C.A., Iravani, R., Kazerani, M., Hajimiragha, A.H., Gomis-Bellmunt, O., Saeedifard, M., Palma-Behnke, R., Jiménez-Estévez, G.A., Hatziargyriou, N.D.: Trends in microgrid control. IEEE Trans. Smart Grid **5**(4), 1905–1919 (2014)
16. Green, T.C., Prodanović, M.: Control of inverter-based micro-grids. Electr. Power Syst. Res. **77**(9), 1204–1213 (2007)
17. Esparcia, J.: Innovation and networks in rural areas. an analysis from European innovative projects. J. Rural Stud. **34**, 1–14 (2014)
18. University of New England, "SMART Farm," (2015). http://www.une.edu.au/about-une/academic-schools/school-of-science-and-technology/research/smart-farm. Accessed 01 June 2015
19. Taylor, K., Griffith, C., Lefort, L., Gaire, R., Compton, M., Wark, T., Lamb, D., Falzon, G., Trotter, M.: Farming the web of things. IEEE Intell. Syst. **28**(6), 12–19 (2013)
20. Hwang, D.-H., Shin, M.-S.: IT convergence technology in plant growing for low-carbon green industry. J. Korea Soc. IT Serv. **11**(4), 123–134 (2012)
21. Scherr, S.J., Shames, S., Friedman, R.: From climate-smart agriculture to climate-smart landscapes, pp. 1–15 (2012)
22. Kenny, G.: Adaptation in agriculture: lessons for resilience from eastern regions of New Zealand. Clim. Change **106**(3), 441–462 (2011)
23. Nadimi, E.S., Søgaard, H.T., Bak, T., Oudshoorn, F.W.: ZigBee-based wireless sensor networks for monitoring animal presence and pasture time in a strip of new grass. Comput. Electron. Agric. **61**(2), 79–87 (2008)
24. Huircán, J.I., Muñoz, C., Young, H., Von Dossow, L., Bustos, J., Vivallo, G., Toneatti, M.: ZigBee-based wireless sensor network localization for cattle monitoring in grazing fields. Comput. Electron. Agric. **74**(2), 258–264 (2010)
25. Ripoll-Bosch, R., Díez-Unquera, B., Ruiz, R., Villalba, D., Molina, E., Joy, M., Olaizola, A., Bernués, A.: An integrated sustainability assessment of mediterranean sheep farms with different degrees of intensification. Agric. Syst. **105**(1), 46–56 (2012)
26. Waller, R.M.: Ground Water and the Rural Homeowner (1994)
27. Konikow, L.F., Bredehoeft, J.D.: Ground-water models cannot be validated. Adv. Water Resour. **15**(1), 75–83 (1992)
28. Walker, G., Devine-Wright, P., Hunter, S., High, H., Evans, B.: Trust and community: exploring the meanings, contexts and dynamics of community renewable energy. Energy Policy **38**(6), 2655–2663 (2010)
29. Bengoa, J.: Historia del pueblo Mapuche, 5th edn. Ediciones Sur, Santiago, Chile (1996)

30. Zhang, X., Chen, Z.D.: The design and simulation of a smart farm system based on ultra narrow band communication. Appl. Mech. Mater. **427–429**, 1398–1401 (2013)
31. Llanos, J., Sáez, D., Palma-Behnke, R., Núñez, A., Jiménez-Estévez, G.: Load profile generator and load forecasting for a renewable based microgrid using self organizing maps and neural networks. In: IEEE World Congress on Computational Intelligence, p. 8 (2012)

Dynamics of the Indices NDVI and GNDVI in a Rice Growing in Its Reproduction Phase from Multi-spectral Aerial Images Taken by Drones

Diego Alejandro García Cárdenas[1(✉)] [iD],
Jacipt Alexander Ramón Valencia[2] [iD],
Diego Fernando Alzate Velásquez[3] [iD],
and Jordi Rafael Palacios Gonzalez[4] [iD]

[1] Instituto Geográfico Agustín Codazzi, San José de Cúcuta, Colombia
diegobioingeniero@gmail.com
[2] Universidad de Pamplona, Pamplona, Colombia
[3] Universidad Santo Tomas, Bogotá, D.C., Colombia
[4] Escuela Colombiana de Ingeniería Julio Garavito, Bogotá, D.C., Colombia

Abstract. In this study, the dynamics of two vegetation indices, the normalized differential vegetative index (NDVI) and the variant of the NDVI that uses the green band (GNDVI) in a rice growing of the variety fedearroz 2000 in reproduction phase, are analyzed. These indices were calculated through the geoprocessing of multi-spectral aerial images taken by a drone or UAVs, with the aim of identifying which zones of the crops are under stress, healthy or dense. The rice growing had an area of approximately 4,1 hectares and its location corresponds to the farm El Faro in the footpath Campo Hermoso within the municipal district of San José de Cúcuta – Norte de Santander. For this research, two flights were carried out, one at the beginning of the reproduction phase dated September 4th 2016 and the second one at the end corresponding to October 8th 2016; these flights were performed with a Iris+ 3DR drone, a canon S100 camera was implemented as a catch images sensor converted into NDVI by using a NGB filter (Near infrared, Green and Blue). As a result, 4 mosaics are shown, one NDVI and one GNDVI on September 4th 2016 and one NDVI and one GNDVI on October 8th 2016, each one of them were classified according to the characteristics observed in field in zones under stress or with low development, healthy and dense zones. Finally, a NDVI dynamic analysis was completed.

Keywords: Precision farming · Drones · Geoprocessing
Multi-spectral images

© Springer Nature Switzerland AG 2019
J. C. Corrales et al. (Eds.): AACC 2018, AISC 893, pp. 106–119, 2019.
https://doi.org/10.1007/978-3-030-04447-3_7

1 Introduction

Rice plays a very important role as a staple food, and the agricultural systems where it is produced are essential for food security, poverty decrease, and the population's lifestyle improvement. Rice became an important agricultural product and an income-generating crop throughout the 20th century. It has evolved from a pioneer crop, mainly rainfed on the agricultural boundary during the first half of that century to become a highly technical and productive crop, where the irrigation system has dominated in recent decades [1]. However, despite the advances, rice, and agriculture in general face phenomena such as variability and climate change that can affect their performance and increase production costs; this generates instability and socioeconomic problems.

In the last decade, climatic variations related to the phenomenon of El Niño and La Niña have brought serious challenges for Colombian agriculture, demonstrating the low capabilities that many farmers have to effectively manage risk and adapt to climatic fluctuations and catastrophes. Anthropogenic climate change is likely to exacerbate this situation. Scientists present improvements in climatic variability, higher temperatures, and erratic precipitation. Probably by the middle of this century, both the average annual temperature and precipitation have increased, which would have significant impacts on agriculture; this has a wide impact on the national economy, rural poverty rates, and food security [2].

Nearby in Norte de Santander, in 2011, climatic variability has impacted rice cultivation; relative humidity averages exceeded 80%, rains were more intense from February to May and November to December. The phytosanitary evaluation of about 90 farms inside and outside the irrigation district of the Zulia River, showed a high incidence of diseases such as helmintosporiosis, sarocladium, and brown spots; these diseases are associated with the stress of the plant. Helmintosporiosis is related to inadequate nutrition plans, nutritional imbalance, nutrient deficiency, toxicity and application of fertilizers out of season [3].

In response to the phenomena of variability and climate change and to cut the negative impacts generated by traditional agriculture to the environment, precision agriculture (PA) is taken as a possible solution to this situation. According to Díaz [4], precision agriculture is the differentiated management of crops using different technological tools (GPS, plant-climate-soil sensors and multispectral images from both satellites and UAS/RPAS). This differentiated management of crop we will be able to detect the variability that has a certain agricultural exploitation, as well as to do an integral management of already stated exploitation.

The current study pretends to give contributions to precision farming through the use of versatile and economic technologies such as drones and unmanned aerial vehicles (UAVs). These are unmanned flight systems, with the capacity of being controlled from land or flying in automatic mode from a geo-referenced flight plan by GPS. They have the capacity of flying at low height and maintaining a real time communication with the station in land [5].

In Norte de Santander, there is a great amount of small crops, with complex fields that cannot be studied from satellite images because their spatial, temporal and spectral resolution does not allow it. Berni et al. [6] enunciate that the current satellites have some limitations for the management of crops, such as the lack of images with optimal spatial and spectral resolutions and unfavorable times of visit to detect the stress in the crops; besides, the study mentions that the manned aerial platforms have high operating costs. The remote detection sensors set in unmanned aerial vehicles can fill this gap, providing low cost focuses to accomplish the critical requirements of spatial, spectral and temporal resolutions. Diaz [4] indicates that the multispectral images of drones capture data from the near infrared and invisible spectra, and depart from them, images of the crop which differentiate the healthy and sick plants can be created. Moreover, it is possible to obtain weekly, daily and even real time information through the drones, which allows to create a temporal series to observe changes and to make appropriate decisions about the crop.

There are studies about drones and multispectral aerial images which have been successfully carried out in agriculture. In the city of Rugao, district of Jiangsu, China, Zhou et al. [7] acquired multispectral and digital images in critical stages of the growth of different rice varieties, they calculated several vegetation indexes, among them are the normalized differential vegetative index (NDVI), and they correlated them with the grain yield and the leave area index (LAI), the results showed that the best stage to predict the performance is the beginning of the panicle (belonging to the reproduction phase), the index with the best results was the $NDVI_{[800.720]}$ which showed a lineal relation with the grain yield and the value of R^2 de 0.75. In Kampung Setia Jaya, Yan, Kedah, Teoh et al. [8] obtained multispectral aerial images, they geo-processed them and conducted a correlation between the performance of the rice and the values of the red bands (R), green ones (G), NRI and NDVI, the correlation showed a $R^2 = 0.748$ between the NDVI and the performance of the crop, this permitted to establish a regression model to estimate the performance of rice from NDVI values. In the Klein-Altendorf campus, 40 kms south of Cologne, Germany, Bendig et al. [9] executed a study of barley to evaluate if the digital models of the crop surface (CSM) obtained from RGB images taken from UAVs can predict the biomass; the comparison between the height of the plant in field (PHref) and the CSM produced a R^2 of 0.92; also, they obtained a high correlation between the CSM and the fresh biomass ($R^2 = 0.81$) and the dry biomass ($R^2 = 0.82$).

The articles deals with dynamic analysis of two vegetative indexes; the normalized differential vegetative index (NDVI) and the NDVI variant that uses the green band (GNDVI) in a rice crop of the variety Fedearroz 2000 in reproduction phase. These indexes were calculated thanks to the geoprocessing of multispectral aerial images taken from drones or UAVs aiming at identifying which zones of the crop are under stress, healthy or dense and analyzing their variability in time and space.

Different from other studies in which satellites to analyze indexes of vegetation or drones with multispectral and hyperspectral camera of high value were employed, this project used two low cost equipment, Canon S100 camera converted to a NDVI through a NGB filter and the Iris+ dron of the 3DR brand. The filter partially blocks the red band (600–675 nm) and allows the entry of a portion of red and near infrared light (676–773 nm), the camera also captures the information from the green bands (441–565 nm) and from the blue ones (384–537 nm). With this capacity for the capture of spectral information, it is possible to estimate several indexes of vegetation, among these are the NDVI and the GNDVI.

2   Methodology

Two flights were developed for this study, one at the beginning of the reproduction phase dated (September 4th 2016) and the second one to at the end (October 8th 2016). This phase goes from the initiation of the panicle to the flowering, it means, it starts when vegetative phase ends. It is characterized by the appearance of the reproductive organs of the plant. The length of it is continuous in all varieties and on average it lasts 35 days [10].

The flights were carried out with the Iris+ 3DR drone, a canon S100 camera was implemented as a catch images sensor converted into NDVI by using a NGB filter (Near infrared – Green and blue). This special filter blocks red light and, in return, it permits the passage of the near infrared above 700 nm. Additionally, it allows the passage of green waves (G) and blue waves (B), thus it has two variants of the visible and one of the near infrared, which enables the basic analysis of the phytosanitary state of the crops: the darker and more intense green color is, the denser and healthier vegetation is [11].

The methodology adopted comprises the following five activities:

2.1   Definition of the Growing and Study Zone

A rice crop of approximately 4.1 hectares was selected which is located in the farm El Faro in the footpath of Campo Hermoso of the district of San José de Cúcuta – Norte de Santander, its geographical location is −72.583857, 8.029046 decimal degrees, in the WGS_1984 geographical coordinate system.

The following images shows the geographical location of the rice growing under study (Fig. 1):

Fig. 1. Location of the growing (Source: Author. Satellite image obtained from Google Earth)

2.2   Design of the Flight Path

The flight path of the Iris+ 3DR drone was designed; in doing so, the Tower app was used. Parameters such as height, flight time and distance covered were taken into account. Due to the growing area, (approximately 5 hectares) and the autonomy of the Iris+ 3DR flight (16–22 min), it was necessary to divide the flight into two parts to avoid possible complications.

The flight height is according to the scale that is pretended to be obtained in the photographs and the focal distance of the camera 5.8 mm. To obtain this resolution of approximately 1.5 cm/pix, a flight height of 50 m was established. At this height, the camera generates a squared image of 47.6 m side; to generate a longitudinal overlap of 60%, it was necessary that the distance between the two expositions or photos were of 19.04 m. The interval of picture taking was obtained thanks to the following formula:

$$I = \frac{B}{V} \tag{1}$$

Where

I = Interval of time among images
B = Distance between two consecutive images
V = speed of the plane.

$$I = 3.7 \text{ s} \tag{2}$$

Characteristics of the flight sections are displayed into the next tables and images (Tables 1 and 2):

Table 1.   Section 1 of the flight

Section 1	
Flying altitude	50 m
Distance	1.4 km
Flight time	4:59 min

Table 2.   Section 2 of the flight

Section 2	
Flying altitude	50 m
Distance	1.9 km
Flight time	6:46 min

2.3   Execution of the Flight and Acquisition of Spatial Data

Flights were executed in automatic mode by a pilot and an observer who supported the mission. For the multi-spectral image capture, a canon S100 camera was implemented as a sensor converted into NDVI.

2.4   Calibration of the NGB Images

For the radiometric calibration, a white piece of foam was used as panel of reference, which was put on the floor during each flight, this served for the correction of the near infrared, green and blue bands of the images obtained by the canon S100 camera.

The following formula was employed for the calibration:

$$\frac{Bx}{P} * SBx \tag{3}$$

Where,

Bx, is the reflectance of each band in the image.
P, is the reflectance of the white piece of foam in the Bx band.
SBx, is the average reflectance of the white piece of foam of the x band measured with a spectroradiometer.

The spectral signature of the white piece of foam was taken using an Apogeo ps-100 spectroradiometer which has a range from 400 nm to 1000 nm, the values of the reflectance in the different wavelength ranges were obtained this way.

2.5   Geoprocessing of Images from the Near Infrared (NGB)

Agisoft Photoscan and Arcgis software was used for geoprocessing of images.
To elaborate orthophoto mosaics, the following steps were established:

Elaboration of NGB Orthophoto Mosaics

- NGB images were aligned using their coordinates into the Agisoft Photoscan software.
- 4 control points were assigned over the NGB images in order to diminish the error in the georeferencing, which permitted to obtain even mosaics. The four points were taken from a Google Earth satellite image. In the following image, the distribution of the control points in the aligned images is detailed (Fig. 2):

Fig. 2. Control points

- The dense cloud of points, the mesh and the texture were created, steps necessary for the working pace of the Agisoft Photoscan in generating the mosaic.
- The NGB orthophoto mosaic was exported as a .tiff file.

Calculation of the Vegetation Indices

In Arcgis software a mesh over the growing was digitalized to generate a 261-point sampling within a distance of 10 m each one; then, a 5-m buffer was created around each point with the aim of analyzing not only a pixel but representative zones.

The normalized differential vegetative index NDVI and its variant of the GNDVI were calculated through the raster calculator of Arcgis software. Afterwards, the zones with values less than 0.2 were excluded, due to the fact that they represent mere soil, rocks and water. Using the buffer and the NDVI and GNDVI mosaics, the tables with the statistics of the 261 buffer for the zone analysis were created through the zone statistics tool.

NDVI Classification

A classification of the orthophoto mosaics was completed according to the different spectral responses. This was based on the comparison of what was observed in field versus the values generated by the NDVI mosaics.

3   Results and Discussions

Two NGB orthophoto mosaics were obtained, one was dated September 4th 2016 and the other one was dated October 8th 2016, with an average height of 50 m, each mosaic geoprocessed with more than 300 images and with a resolution of approximately 1.5 cm/pixel. The characteristics of the mosaics are listed below (Tables 3 and 4):

Table 3. Characteristics of the NGB orthophoto mosaic (September 4th 2016)

Characteristic	Value
Number of images	352
Flying altitude	49.14 m
Ground resolution:	0.0151347 m/pix

Table 4. Characteristics of the NGB orthophoto mosaic (October 8th 2016)

Characteristic	Value
Number of images	339
Flying altitude	52.16 m
Ground resolution:	0.0134423 m/pix

From this NGB orthophoto mosaics, the following four mosaics were generated:

- By the date of the beginning of the reproduction phase September 4th 2016, 1 NDVI mosaic and 1 GNDVI were obtained.
- By the date of the end of the reproduction phase October 8th 2016, 1 NDVI mosaic and 1 GNDVI were obtained.

The pixels containing NDVI values less than 0.2 were eliminated, it is why in NDVI mosaics it is observed blank or without data spaces, these were furrows of land disposed to watering system, as observed in the following image (Fig. 3):

Fig. 3. Aerial photograph of the growing

3.1   NDVI Classification

Mosaics were classified by ranks of values of NDVI in order to visualize different zones of spectral response as follows (Table 5):

Table 5. NDVI classification of the orthophoto mosaics

Id	NDVI	Color	Characteristic
1	0.2 to 0.3	Orange	Vegetation under stress or few developed
2	0.3 to 0.6	Light green	Healthy vegetation
3	0.6 to 1	Dark green	Dense vegetation

This classification was based on the comparison of what was observed in field versus the values of the NDVI mosaics, also what was consigned in [11, 12].

The zones under stress, the healthy and dense zones already defined are associated to the spectral response of the rice plant, because the NDVI values are established depending on the energy absorbed or reflected by plants in several parts of the electromagnetic spectrum. The spectral response of the healthy vegetation shows a clear contrast between the spectrum of the visible one, specially the red band (for this study the blue band was employed), and the Near Infrared (NIR) [4].

The previous statements are supported by many authors; for instance Araque and Jimenez [13] state: the region of the visible spectrum in the vegetation is characterized by low reflectance and transmittance, due to the strong absorption by leaf pigments. For example, chlorophyll pigments absorb the violet-blue and red light because of photosynthesis. The green light is reflected for the photosynthesis, reason why many plants emerge green.

Sanger [14] enounces that when the leaf is sick, the chlorophyll degrades faster than carotenes. This effect causes a rise on the reflectance of the red wavelength, owing to the reduction of the absorption of the chlorophyll. Carotenes and Xanthophylls are now dominant in leaves, and the leaves appear being yellow because of carotenes and xanthophylls absorb blue light and reflect green and red light.

Abdullah and Umer [15] propose that damages by diseases and plagues can be measurable according to variants in the content of the plants chlorophyll, which can be analyzed in terms of their changes in their spectral images patterns taken by satellites. These techniques use multi-spectral images to detect areas under stress.

3.2   Dynamic Analysis NDVI–GNDVI

The NDVI mosaics of the beginning (September 4th 2016) and end dates (October 8th 2016) of the reproduction phase are shown below (Fig. 4):

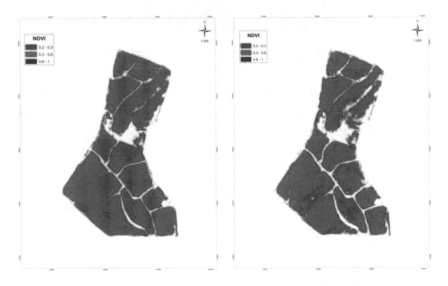

Fig. 4.   September 4th 2016 NDVI (left) and October 8th 2016 NDVI (right)

The NDVI mosaic sampling dated September 4th 2016 established that the minimum value was 0.29, the maximum one was 0.54 and the average was 0.40, whereas the sampling dated October 8th 2016 yielded a minimum value of 0.42, a maximum of 0.61 and an average of 0.52. An increase in the NDVI value was observed over a period of roughly 35 days (period belonging to the reproductive phase of cultivation), the underdeveloped or stressed areas (which were minimal) became healthy zones and an important area of healthy zones managed to be dense. This NDVI growth is due to the fact that the cultivation was in the reproductive phase, which includes the initiation period of the panicle until the flowering, the panicle primordium develops and the size of the plant and dry matter increase rapidly, the height of the plant shows a progressive growth up to the flowering (98 days of development), while the leaf area increases until reaching its maximum value at 91 days. The phase is characterized by the elongation of the stem, emergence of the decayed leaves, stuffing and filling of the spikelets, it lasts until maturation and it is marked by an increase in the weight of the panicle accompanied by a decrease in the weight of the straw [16].

The increase in the NDVI value and therefore in the rice crop growth coincides with the study conducted by Zhou et al. [7], in which its correlation between the performance of the rice and the $\Sigma NDVI_{[800.720]}$ had a $R^2 = 0.75$; this study established that the best period for the calculation of the rice performance is the starting or beginning stage of the panicle, due to the production of yellow leaves raises the difficulty of calculation of the LAI and the prediction of the performance in the last growth stages.

For the analysis of the results of the date September 4th 2016, the NDVI and GNDVI mosaics are shown as follows (Fig. 5):

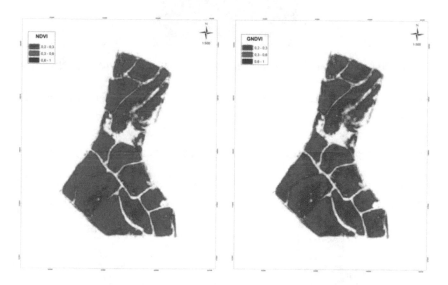

Fig. 5. September 4th 2016 NDVI (left) and September 4th 2016 GNDVI (right)

The GNDVI is calculated similarly to the NDVI, but the green band is used instead of the red band. It is related to the radiation proportion absorbed photosynthetically and is linearly correlated with the leaf area index (LAI) and the base. Therefore, GNDVI is more sensitive to chlorophyll concentration than NDVI and varies from 0 to 1.0 [17].

In the NDVI mosaic dated September 4th 2016 (left image), the crop presented healthy vegetation (light green) in most of its extension, followed by zones with stressed vegetation in minor proportion and dense zones were absent. At this time the cultivation was finishing the vegetative phase and began the reproductive phase of its development. The vegetative phase of delay, maximum tillering, internode elongation and panicle initiation occur almost simultaneously, the size of the seedling and the increase of the dry matter increase at a lower speed and the number of culms decreases. Its duration depends on the variety and climatic conditions, especially the duration of the day and the temperature [16].

In the following image the development of the rice plant in this phase is shown (Fig. 6):

Fig. 6. Rice plant at the beginning of the reproduction phase (September 4th 2016)

For the analysis of the results on October 8th 2016, the following NDVI and GNDVI mosaics are shown (Fig. 7):

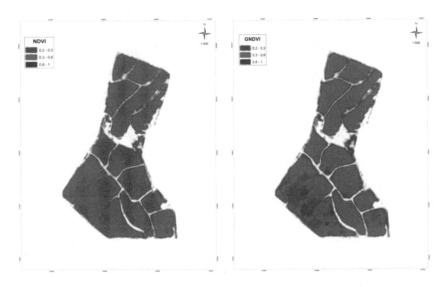

Fig. 7. October 8th 2016 NDVI (left) and October 8th 2016 GNDVI (right)

During this time the growing showed NDVI values usually above 0.4 with a large number of healthy and dense areas (left image). Few peripheral zones emerged with vegetation under stress, this vegetation belongs to weeds, according to what was observed in field, as shown in the following image (Fig. 8):

Fig. 8. Weeds in the periphery of the growing

In this date, the growing was in the flowering stage, which occurs from the emerging of the panicle of the flag leaf sheath, until when the anthesis is completed throughout the panicle. It has a lapse from 7 to 10 days [10].

In the comparison between NDVI mosaics versus GNDVI mosaics, a difference of the average values of the zones was noticed which is comprised from 0.02 to 0.16 on the first date (September 4th 2016) and from 0.10 to 0.22 on the second date (October 8th 2016); generally, in both cases, the zones where NDVI increased, GNDVI did it as well and viceversa. Furthermore, as GNDVI values were from 0.02 to 0.2 below NDVI, some parts of the GNDVI mosaic showed zone with a lower classification according to NDVI parts, it means, where NDVI mosaics registered healthy zones, GNDVI evidenced zones under stress. Finally, it is important to mention that this project and classification were emphasized on NDVI index, which is the most cited index for these kinds of analysis.

4   Conclusions

Through the Iris+ 3DR drone with a canon S100 camera converted to NDVI, it is possible to collect information from the visible spectrum bands and NIR with a good real-time spatial and spectral resolution of agricultural crops (in this case of rice), from there to generate orthophoto mosaics of vegetation indices (NDVI - GNDVI), this allows to analyze different areas of spectral response and to make precise decisions in order to combat the spread of diseases and plagues, to intervene zones with nutritional deficiencies, to decrease chemical products dependency, costs and negative impacts towards the environment.

The NDVI varied in the rank from −1 to 1, the zones belonging to mere soil, rocks and water generated values lower than 0.2, the vegetation under stress was in the rank from 0.2 to 0.3, healthy plants showed values from 0.3 to 0.6 and finally the dense vegetation showed values above 0.6. This demonstrates that healthy plants generate a high contrast between the blue and the near infrared bands, used for the calculation of the NDVI, while in the underdeveloped or under stress is minimal. Therefore, it can be deduced that healthy plants pigments such as chlorophyll and carotenoids of the photosynthetically active region (visible spectrum) have a high absorption of energy and reflect a minimal part, whilst in the infrared the reflected energy is high.

For the rice producers, the spatio-temporal statistics of the vegetation indexes obtained through multispectral cameras installed in drones become a fundamental tool to undertake sustainable agriculture when acquiring detailed information and, in early stages of the rice crop, problems, diseases or deficiencies can be detected so that an adequate management of all processes can be made such as: applying fertilizers just on necessary zones, diminishing the application of chemical products (herbicides, pesticides, insecticides and fungicides) and estimating the performance of the rice grain in order to undertake actions that increase the production and decrease the costs.

The classification of the healthy and sick zones and the analysis of the vegetation indexes allow the grower to conduct a correct management of the water, doing the watering in zones that are really damaged, this appears to be relevant in dry seasons and attenuates the impact by reduction of the hydrological resource.

References

1. Sanint, L.: Nuevos retos y grandes oportunidades tecnológicas para los sistemas arroceros: Producción, seguridad alimentaria y disminución de la pobreza en América Latina y el Caribe. In: Degiovani, V., Martínez, C., Motta, F. (eds.) Producción ecoeficiente del arroz en América Latina, vol. 370, pp. 3–12. Centro internacional de agricultura tropical (CIAT), Cali, Colombia (2010)
2. Lau, C., Jarvis, A., Ramírez, J.: Agricultura Colombiana: Adaptación al Cambio Climático. In: CIAT Políticas en Síntesis No. 1, pp. 1–4. Centro Internacional de Agricultura Tropical (CIAT), Cali, Colombia (2011)
3. Cuevas, A.: El clima y el cultivo del arroz en Norte de Santander. Revista arroz 60(497), 4–8 (2012)
4. Díaz, J.: Estudio de índices de vegetación a partir de imágenes aéreas tomadas desde UAS/RPAS y aplicaciones de estos a la agricultura de precisión. Universidad Complutense de Madrid, Madrid, España (2015). https://eprints.ucm.es/31423/1/TFM_Juan_Diaz_Cervignon.pdf
5. Fajardo, J.C.: Apoyo a la agricultura de precisión en Colombia a partir de imágenes adquiridas desde vehículos aéreos no tripulados (UAV's). Universidad Javeriana, Bogotá D. C., Colombia (2014). https://repository.javeriana.edu.co/bitstream/handle/10554/16484/FajardoJuncoJuanCamilo2014.pdf?sequence=1
6. Berni, J., Zarco-Tejada, P., Suárez, L., Fereres, E.: Thermal and narrowband multispectral remote sensing for vegetation monitoring from an unmanned aerial vehicle. IEEE Trans. Geosci. Remote Sens. 47, 722–738 (2009). https://doi.org/10.1109/TGRS.2008.2010457

7. Zhou, X., Zheng, H.B., Xu, X.Q., He, J.Y., Ge, X.K., Yao, X., Cheng, T., Zhu Y., Cao, W. X., Tian, Y.C.: Predicting grain yield in rice using multi-temporal vegetation indices from UAV-based multispectral and digital imagery. Photogramm. Remote. Sens. **130**, 246–255 (2017). https://doi.org/10.1016/j.isprsjprs.2017.05.003

8. Teoh, C., Mohd Nadzim, N., Mohd Shahmihaizan, M., Mohd Khairil Izani, I., Faizal, K., Mohd Shukry, H.: Rice yield estimation using below cloud remote sensing images acquired by unmanned airborne vehicle system. Int. J. Adv. Sci. Eng. Inf. Technol. **6**(4), 516–519 (2016). http://dx.doi.org/10.18517/ijaseit.6.4.898

9. Bendig, J., Bolten, A., Bennertz, S., Broscheit, J., Eichfuss, S., Bareth, G.: Estimating biomass of barley using crop surface models (CSMs) derived from UAV-based RGB imaging. Remote Sens. **6**, 10395–10412 (2014). https://doi.org/10.3390/rs61110395

10. Moquete, C.: Guía técnica el cultivo de arroz. In: Serie Cultivos, No. 37, pp. 1–166. CEDAF, Santo Domingo, República Dominicana (2010). http://www.cedaf.org.do/publicaciones/guias/download/arroz.pdf

11. Berrio, V., Mosquera, J., Alzate, D.: Uso de drones para el análisis de imágenes multiespectrales en agricultura de precisión. Ciencia y Tecnología Alimentaria, **13**(1), 28–40 (2015). https://doi.org/10.24054/16927125.v1.n1.2015.1647

12. Esri: Función NDVI. https://pro.arcgis.com/es/pro-app/help/data/imagery/ndvi-function.htm. Accessed 07 June 2018

13. Araque, L., Jiménez, A.: Caracterización de firma espectral a partir de sensores remotos para el manejo de sanidad vegetal en el cultivo de palma de aceite. Revista Palmas 30(3), 63–79 (2009)

14. Sanger, J.E.: Quantitative investigation of leaf pigments from their inception in buds through autumn coloration to decomposition in falling leaves. Ecology **52**, 1075–1089 (1971)

15. Abdullah, A., Umer, M.: Applications of remote sensing in pest scouting: evaluating options and exploring possibilities. In: Proceedings of 7th ICPA, Julio 25–28, Minneapolis, MN, USA (2004)

16. Degiovanni, V., Gómez, J., Sierra, J.: Análisis de crecimiento y etapas de desarrollo de tres variedades de arroz (Oryza sativa L.) en Montería, Córdoba. Temas Agrarios, **9**(1), 21–29 (2004). https://doi.org/10.21897/rta.v9i1.620

17. Candiago, S., Remondino, F., De Giglio, M., Dubbini, M., Gattelli, M.: Evaluating multispectral images and vegetation indices for precision farming applications from UAV images. Remote Sens. **7**(4), 4026–4047 (2015). https://doi.org/10.3390/rs70404026

Wireless Sensor Network for Monitoring Climatic Variables and Greenhouse Gases in a Sugarcane Crop

Oscar A. Orozco[1,2]([✉]) [iD] and Gonzalo Llano[1] [iD]

[1] Universidad Icesi, Cali Valle 760031, Colombia
oaorozco@icesi.edu.co
[2] LiveVox Inc., Medellin Antioquia 050022, Colombia

Abstract. In Colombia, sugarcane is one of the most important crops with about 235,000 hectares cultivated at the end of 2017. Therefore, establishing and implementing research projects pursuing objectives as the design of agricultural production procedures which allow more productivity, cleanness, and efficient processes are welcome. We present a wireless sensor network proposal oriented to deal with the lack of monitoring in some agronomic/climatic variables. The prototype is intended to measure variables such as temperature and soil/air moisture. Also measured are soil pH and the most relevant greenhouse gases, i.e., carbon dioxide, methane, and nitrous oxide. We focused our work in measuring the greenhouse gases to analyze the values in parts per million, seeking to establish a reference point to potentially calculate the carbon footprint associated to the sugarcane cultivation.

The closed-chamber technique was used for the measurements and the initial results showed that the gases concentrations were higher outside the chamber: 376 ppm of CO_2, 1.31 ppm of CH_4, and 0.504 ppm of N_2O (outside) compared with the values inside them: 291 ppm of CO_2, 1.01 ppm of CH_4, and 0.163 ppm of N_2O (inside) as average values. Nevertheless, bias was observed during the measurements due to the high current demand the gas sensors needed. We expect to solve these issues and increase the amount of measurements for upcoming releases of the presented prototype.

Keywords: Carbon footprint · Greenhouse gas · Open source hardware
Sugarcane · Wireless sensor network

1 Introduction

Agriculture is a task carried out by humankind thousands of years ago and its processes have evolved through time. Currently, citing data of the Food and Agriculture Organization of the United Nations (FAO), almost one third of the global population depend on the agriculture for their maintenance and in emergent economies, this sector can represent up to 30% of the gross domestic product [1]. Consequently, several governments and institutions (both public and private) have started projects with the aim of improving the agricultural sector in these countries. In Colombia, the national development plan empowers the transformation of the rural areas pursuing a sustainable

© Springer Nature Switzerland AG 2019
J. C. Corrales et al. (Eds.): AACC 2018, AISC 893, pp. 120–135, 2019.
https://doi.org/10.1007/978-3-030-04447-3_8

growing through the modernization of processes, products, and services. Hence, the Information and Communication Technologies (ICT) have started to be applied into the agricultural sector to enhance the general yield of the crops in the scope of the so-called Precision Agriculture (PA).

The PA can be defined as the act of monitoring, controlling, assessing the right nutrients, and apply them given the necessities of the studied soil and the field variability and site-specific conditions. Moreover, the technologies used for these purposes are as variable as the Global Positioning Systems (GPS), Geographic Information Systems (GIS), yield monitors, remote sensors, and variable-rate applicators [2]. In a country like Colombia where most of the agriculture actors still use traditional activities and the PA features are still unknown, novel solutions based on research projects to enhance monitoring activities are welcomed. Sugarcane farmers —most of the times— use empiric methods to apply fertilizers, water, and nutrients to the crop and they do not have tools to be informed of the actual status of determined elements that, potentially, could increase the crop yield. Besides, it is not a secret that the emission of greenhouse gases (GHG) tends to grow and the trend is to consider actions to reduce the carbon footprint associated to this rise.

Because of the potential that sugarcane has in the Colombian economy, solutions that facilitate the monitoring of the main agronomic (pH, soil moisture) and climatic (temperature, moisture, light intensity) variables altogether with the main GHG are important to increase the efficiency in the production processes. Furthermore, the quantification process of the potential crop carbon footprint (i.e., the contamination generated by the procedures carried out when the crop inputs are applied) is also a general concern of many companies nowadays. For this reason, the main motivation of this research project entails the implementation of a Wireless Sensor Network (WSN) to monitor the main agronomic and climatic variables of the sugarcane together with the potential emissions of the main GHG (carbon dioxide, CO_2; methane, CH_4; and nitrous oxide, N_2O). We will perform a comparative study of the values collected with the WSN with values analysed in a professional laboratory to assess the reliability of our network. The procedure to collect these samples for the laboratory was performed in parallel with the measurements we recorded with the sensors.

Considering this, we expect to answer the following research question: is it viable to implement a WSN capable to monitor the agronomic and climatic variables of an organic sugarcane crop, plus the GHG emissions associated to the mechanical labours performed within it?

Our proposal tries to address the research question by using low-cost technologies that require basic knowledge of certain technologic topics and that does not come to replace the farmer, but to complement his/her labours.

2 Materials and Methods

For the execution of our proposal, we start by presenting a summary of the reachable scope of the prototype, followed by some data of the studied area, the description of the elements conforming all the WSN (i.e., the architecture), and concluding with the employed methodology.

2.1 Scope and Data of the Studied Area

Pursuing the answer of the research question previously mentioned, the scope of the proposal presented in this article is to implement a WSN to monitor several variables in an organic sugarcane crop. This implementation is intended to measure the concentration of the most relevant GHG using the closed-chamber technique (displayed in Fig. 1) and compare the results with gas samples taken from the same chambers and sent to be analyzed in a special laboratory. Details of this procedure are described by Lubo et al. [3] and the authors helped in the gathering of those samples while they took measurements with the sensor network described in this document.

Fig. 1. Chamber used in this project and in [3]

Regarding the studied crop, we can mention that the sugarcane belongs to the *Poaceae* family —like rice— and its cultivars are clones that are propagated by stem cuttings [4]. Its main product is the sugar, even though there are other by-products like vinegar, carburant alcohol, panela, paper pulp, vinasse, and bioethanol [5]. In Colombia, most of the sugarcane cultivations are in the geographic valley of the Cauca river, located in the southwest part of the country. It entails 47 towns with a total area of 208,254 hectares, annual precipitations of 1000 mm/year, and an average temperature of 25 °C [6].

The chosen sugarcane lot for the implementation of the WSN is in the *Bugalagrande* town, within the Cauca river geographic valley. The farm conserves traditional labors related with the sugarcane plantation and the lots have conservation agriculture features. Other features of the sugarcane plantations in the farm are the use of organic fertilizers (i.e. chicken manure, poultry litter, and sheep manure), weed control with lambs, minimal mechanical soil tillage (only once during the crop life-time), controlled scarification, intercropping with cowpea (*Vigna unguiculata*) as a natural nitrogen fixer, and the avoidance of the burning process.

2.2 Architecture of the WSN

The general architecture of the implemented WSN was planned considering five sensor nodes placed next to the chambers of the parallel process. These sensor nodes are composed of the open source board, the sensors themselves, the proprietary wireless transceiver, and the power section (solar panel and lithium polymer battery). These elements were in the sugarcane crop to measure the desired variables. In contrast, it was planned that the sensors would be supported by two master nodes, capable to retransmit the information to the farmhouse. The transceivers use the IEEE 802.15.4 set of standards to communicate each other and they operate in the 2.4 GHz band.

Nevertheless, for the results presented in this document, we only implemented a communication scenario between the wireless transceiver in two chambers and a laptop computer with a transceiver plugged via USB in the sugarcane field. The main reason of this was the maximum distance reached by the wireless modules; for this prototype, the wireless modules had a maximum reach of only 25 m and 150 m. The repeater function is pending to be implemented in the following stage of the project via a RF transceiver operating at 427 MHz and with an external directive antenna. The chosen network topology was the star-tree one, since the information between the sensor nodes is sent to the master ones through a single link. The nodes were set into the Application Programming Interface (API) mode, since they send the information to the master nodes in categorized frames, allowing a better understanding in the reception node of the data via the Received Signal Strength Indicator (RSSI) per packet. Besides, the API mode allowed us to receive an ACK of each sent packet.

Components and General Operation. The main components of the WSN were the development boards, wireless modules, solar panels, batteries, and sensors. They were all encapsulated to operate as a single mobile node. A description of the main subsystems conforming the WSN is presented below.

Development Board. From the collection of development boards commercially available, we selected one with open source specifications, since we consider it satisfies all the requirements related with sensors and shields compatibility, processing power, reduced size, and operation with LiPo batteries and solar panels. We discarded other development boards since their cost is considerably higher than the selected one, even though if they are more powerful. However, the computational performance provided by the used one is sufficient to the scope of this proposal.

Sensors. The main sensors for this project were the gas ones measuring CH_4, N_2O, and CO_2. For the encapsulation and posterior burying of the sensors it was necessary to put every component —except the solar panel and the sensors— in a closed pan to avoid issues after their implementation in the sugarcane field trial. Also, we took measurements inside and outside the chamber, so the pans in the exterior were sealed with liquid glue in the top. Figure 2 presents both the pan with the sensors inside and outside the chamber; here, the reader might notice that the sensors were left outside the pans and the solar panel was located on top of it.

As per the datasheet of the sensors, they need a preheat time to achieve adequate results. Hence, for this prototype, we considered a preheat time of 12 h with a calibration period in a closed chamber because we had some time limitations relative to the moment where the sensors were put in place. For this reason, the future prototypes should deal with the preheating time in a closed and controlled environment; we plan to address this issue with sensors not requiring a large preheating time or increasing the batteries capacity to increase the system standby time.

Fig. 2. Encapsulation of the sensors within the pans

Wireless Communication. In the wireless topic, we decided to use proprietary transceivers for the sensor nodes. This was done pursuing, reliability, portability, and ease to use. The sensor nodes next to the chambers mounted a transceiver with a small-integrated dipole antenna. It is important to focus that the studied area consists of 1 hectare of organic sugarcane, where the sensors were placed next to the chambers of the parallel process. The total distance between the sensors location and the farmhouse is 2.26 km in line of sight. Consequently, the need of a repeater in the middle of this line of sight path is necessary to assure the correct communication between the sensor nodes and the workstation in the farmhouse. Hence, we plan that the repeaters should use an RF module operating at 427 MHz and with an external high-gain antenna, capable to reach greater distances. The transceiver reference is HC-12 and it is planned to be used with a UT-106UV antenna. Although, this phase of the project was not covered in this prototype presented in the current document.

Solar Panels and Batteries. The described proposal takes energy from LiPo batteries with 1000 and 2000 mAh of nominal capacity. Also, the solar panels used to charge the batteries provided up to 1 W of power with dimensions of 80 × 100 mm. Besides, for the correct operation of this subsystem, we used a specialized energy board to allow a constant charge to the battery.

2.3 Methodology

For this proposal and after the research carried out regarding the topic of interest, the methodology employed was a combination of steps from other methodologies. Punctually, the feature to finish tasks in a sequential way is inherited from the waterfall model [7], the hardware design process is taken and customized from the work presented by Zeidmann [8], and the scenario description is considered from the steps described by Ulgen et al. [9]. Figure 3 presents the steps in the employed methodology to successfully achieve the proposed objectives.

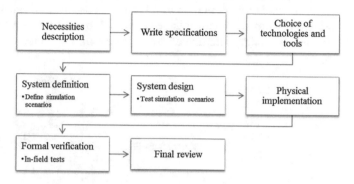

Fig. 3. Employed methodology for the project

As mentioned before, the phases in the methodology are sequential, but there are no restrictions whether a phase can start earlier (i.e. the labors might be carried out several times at the same time). The measurement process was performed next to each chamber, in 30 s intervals for each sensor reading. This first prototype was focused on the collecting of the data related with the greenhouse gases (i.e., CO_2, N_2O, and CH_4). This, trying to verify if those data are related with the ones captured and analyzed in a specialized laboratory.

Also, it is important to mention that the gas samples sent to analysis in a specialized laboratory were taken simultaneously with the sensor readings. Further details of the procedure followed for this can be found in the work published by the authors and other colleagues, described in [3].

2.4   System Design Parameters

After the description of the elements conforming the WSN, the main design patterns used to build and interconnect all the elements are described in the following sections, starting with the choice of topologies (physical and logical) and the most relevant parts of the code. The design pattern term was adjusted from the software engineering area to adapt it to the work performed in this document to let the reader know that the code, schematics, and elements used in the WSN are characterized to have template features for future updates and these templates work with the used elements. Shalloway and Trott [10] define a design pattern as a general repeatable solution to a commonly problem occurring when designing software. This solution is not the final implementation of a project, but an element that can be transformed searching the final solution for the client. That approach is the desired one in this work, since one of the ultimate goals is to provide information for future projects and to be a basis with the code, schematics, and architecture proposed.

Topologies. The employed topology for the WSN was a hybrid between the star and tree ones, since the sensor nodes in the field collect the information and send it to the master and, eventually, this master node sends the data to the repeater. This latter resends the data to the final receiver in the farm. Figure 4 presents the logical topology (i.e., the operational one) of the nodes, where a similitude with the tree topology is observed, but the nodes sending information to a single sink makes the difference.

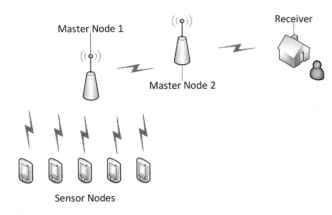

Fig. 4. Logical topology of the WSN

For the physical topology, the differences are not considerable relative to the logical one. The main modifications are related with the exact location the nodes were placed in the sugarcane studied parcel. Considering that the parcel has 1 hectare and its shape is almost a square, the position of the sensor nodes was in the diagonal of the square; whilst the location of the master node within the sugarcane was near one of the corners of the imaginary square. This was performed to ensure the correct communication between the sensor and master nodes. The master node 2 in Fig. 4 was placed on the top of a tree located at ∼1 km in Line of Sight (LOS) of the farthest sensor and, at the same time, this latter is located at 1.3 km of the receiver position. On the other hand, the sensors were located at 30 m approximately each from the other ones to ensure the correct communication of the weakest transceiver devices. Figure 5 presents a schematic of the network physical topology. Here, it is important to point that, due to the fact the sugarcane was on its latest growing stage and its height was up to 2 m, the fading of the wireless signals was considerable. Hence, some packets were lost during the communication, but nothing severe to affect the transmission in a constant way.

Logic in the Code. As per the use of the open source hardware boards to build the sensor nodes, the programming of the boards and the sensors was made using the dedicated Integrated Development Environment (IDE) of the board, which is a software application written in Java and it is based on the Processing and Wiring programming languages. The codes derived from this IDE are called sketches with "*.ino*" file extension and they are characterized by having two well-defined parts: the starting of it and the main loop. In the start, the variables initialization is performed, whilst in the loop the repetitive operations are placed. Alternatively, the support for libraries is

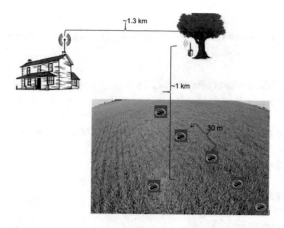

Fig. 5. Schematic of the physical topology of the WSN

extensive and updated, there is an active community of developers and enthusiasts making projects with the employed board —which eases the code reuse—, and the IDE is multi-platform [11].

The first part of the codes uploaded into the development boards is related with the variables initialization and selection of analog ports to be used. Then, in the loop part, the measurements are taken with a delay of 5 s per sample and per sensor; that is, the system starts with the MQ4 gas sensor, waits 5 s and takes the measurement of the MQ135. It waits another 5 s and then takes the measurement of the moisture sensor, then the water drops sensor, and finally the pH sensor; this was done to avoid the overloading of the sensors and the board and to allow a correct transmission of the information reducing the collision probability. Furthermore, it is important to emphasize the code parts where the measurement of the moisture and pH variables is performed.

The moisture sensor presents three results: it detects if there is water in the soil and if the sensor is immersed in water. These values are taken by the analogue measurement of it. On the other hand, the listing below presents the part where the pH sensor measurement is displayed on screen. From here, the first part of the code samples the sensor measurements into a vector to smooth the values, then organizes those values from small to large, convert the values to millivolts, and then to the pH value.

These conversions are required to achieve a more accurate pH measurement from the sensor. As mentioned before, every 5 s a different sensor sends its results to avoid network congestion and a correct visualization of the ACKs.

```
for(int i=0;i<10;i++)     //Get 10 sample values from the sensor
for smooth the value
  {
    buf[i]=analogRead(pHsensor);
    delay(10);
  }
  for(int i=0;i<9;i++)     //sort the analog from small to large
  {
    for(int j=i+1;j<10;j++)
    {
      if(buf[i]>buf[j])
      {
        temp=buf[i];
        buf[i]=buf[j];
        buf[j]=temp;
      }
    }
  }
  avgValue=0;
  for(int i=2;i<8;i++)     //take the average value of 6 center
sample
    avgValue+=buf[i];
  float phValue=(float)avgValue*5.0/1024/6; //convert the analog
into millivolt
  phValue=3.5*phValue;      //convert the millivolt into pH value
  Serial.print("    pH,");
  Serial.print(phValue,2);
  Serial.println(" ");
  delay(5000);
```

[Logic of the pH sensor in the sketch. By the Authors]

As per the work described by Vaduva [12], the MQ-X family of sensors require calibration to achieve accurate values of the measured gas. For that reason, a calibration time of 12 h was employed due to time limitations (the sampling day was too near of the time the sensors were built); besides, the author also suggests the minimal current these sensors should handle (150 milliamperes (mA) typical, 250 mA suggested). This fact was a limiting of the results described in the next chapter and the reader can find all the details there.

The gas sensors operate by measuring the modification of a built-in resistor according to the concentration of the gas; i.e., if the concentration is high, the resistance decreases and vice versa. Besides, they have a variable load resistor that for this case was set up to 4.4 kΩ. It is necessary to calculate R_o as the resistance of the sensor in a common outdoor scenario. The following listing shows the code for this.

```
float sensor_volt; //Define variable for sensor voltage
  float RS_air; //Define variable for sensor resistance
  float R0;
  float sensorValue; //Define variable for analog readings
  for (int x = 0 ; x < 500 ; x++)
  {
    sensorValue = sensorValue + analogRead(A0); //Add analog
values of sensor 500 times
  }
  sensorValue = sensorValue / 500.0; //Take average of readings
  sensor_volt = sensorValue * (5.0 / 1023.0); //Convert average
to voltage
  RS_air = ((5.0 * 10.0) / sensor_volt) - 10.0; //Calculate RS
in fresh air
  R0 = RS_air / 4.4; //Calculate R0
```

[Calculation of the R_o internal sensor resistance. By the Authors]

Consequently, this R_o value was calculated for the employed MQ-X sensors employed in this project. Then, as per Castro [13], obtained values from the sensors have to be multiplied by 2 and divided by 1000 (considering its own R_o). These results are further explained in the next section.

3 Results and Discussion

Considering that the communication was performed between the transceivers in two chambers and a laptop computer receiving the data, it is important to focus that, at the moment of the sampling, the sugarcane height was between 2 and 2.5 m (it was ready to cut). Hence, the wireless connectivity tests showed that the maximum distance reachable for the wireless modules was up to 75 m. For this reason, the wireless performance of the system was not assessed in this document. It is pending to implement the sensor employing an HC-12 transceiver operating at 433 MHz with a Nagoya antenna to allow the reception of the data in the farmhouse, located at 2.26 km in line of sight from the place the chambers were.

It is important to focus that this prototype was tested only during one of the latest visits to the sugarcane lot and the taken measurements were up to 1 h because the parallel study (the one with manual gas gathering and posterior analysis) was set to last that time. For future releases, we plan to extend the measurement time.

3.1 Results

The first data captured by the sensors presented a similar behavior than the atmospheric levels of the assessed greenhouse gases. All the following measurements were taken on December 1st, 2017 for two chambers. Table 1 presents the average concentrations for these gases inside two chambers on each sampling time (i.e., each 20 min).

Table 1. Average sensor values of greenhouse gases inside the chambers

Chamber 1

Sampling time	Temperature (°C)	CH$_4$ value (ppm)	N$_2$O value (ppm)	CO$_2$ value (ppm)
7:16 to 7:36	21.6	1.58	0.546	386
7:36 to 7:56	21.7	1.35	0.103	276
7:56 to 8:16	22.0	1.29	0.101	254
8:16 to 8:36	22.2	0.84	0.100	250

Chamber 2

7:16 to 7:36	21.8	1.15	0.592	306
7:36 to 7:56	22.6	1.11	0.610	298
7:56 to 8:16	23.2	1.42	0.648	264
8:16 to 8:36	23.6	1.83	0.777	245

In this study, we only took measurements with the sensors in one occasion; hence, the data mentioned here was the one to be treated and analyzed. We expect to increase the number of measurements to gather more data and study it with mean and standard deviation values. But for the case of the results presented in this article, the data to be considered is the one taken the day the measurements were performed.

From the table above, it is possible to see that the trend of all the three assessed gases —on average— in the chamber 1 is to reduce over time. Considering that the chamber is always sealed, it is possible that this behavior is caused by the lack of interaction between the gas particles. Even though the results in the chamber 2 differ from the first one in the CH$_4$ and N$_2$O levels, the CO$_2$ concentrations always presented a trend to diminish. From here, the reader can observe that the concentrations in the chambers are near of the atmospheric levels, a fact that will be interesting to contrast in the following subsection. The temperature values have the trend to increase, since the atmospheric conditions in the day of the sampling were clear. Hence, the sunrise made the temperature to increase a few degrees inside the chambers.

We also took some measurements outside the chambers to compare the results with the ones obtained with the Lutron GCH-2018 m, a device capable to measure the CO$_2$ concentration in a specific area. Figure 6 presents the CO$_2$ concentration and temperature for the time of the sampling —i.e., between 7:15 a.m. and 8:45 a.m. (1 h and 30 min)— using this Lutron element. It is important to focus that this element takes a sample of the data every 8 s, where the maximum CO$_2$ value was 504 ppm, the minimum was 281 ppm, and the mean was 366 ppm.

Likewise, Table 2 presents the sensors data for the three assessed gases outside the chambers, where the concentrations are presented in ppm. The concentrations were higher outside than inside the chambers. This can be explained by the fact that the gas values can be affected by several variables as wind, temperature, humidity, and artificial CO$_2$ sources near the measurement point.

Comparing the CO$_2$ values obtained from the Lutron unit (Fig. 6) with the ones in Table 2, the reader might infer that the range of the sensors is between the maximum, minimum, and average values obtained from the element. Nevertheless, the data in Fig. 6 was sampled each 8 s, which provides a higher precision than the one obtained

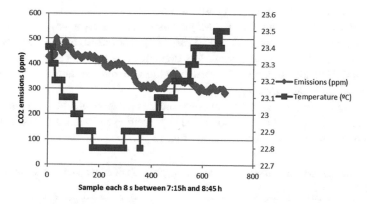

Fig. 6. CO_2 and temperature values from the Lutron meter

Table 2. Average sensor values of greenhouse gases outside the chambers

Chamber 1				
Sampling time	Temperature (°C)	CH_4 value (ppm)	N_2O value (ppm)	CO_2 value (ppm)
7:16 to 7:36	21.6	1.02	0.419	436
7:36 to 7:56	21.7	1.02	0.606	530
7:56 to 8:16	22.0	1.01	0.614	536
8:16 to 8:36	22.2	1.02	0.560	495
Chamber 2				
7:16 to 7:36	21.8	1.78	0.650	265
7:36 to 7:56	22.6	1.26	0.680	250
7:56 to 8:16	23.2	1.40	0.640	267
8:16 to 8:36	23.6	1.30	0.600	299

with the sensors (sampled approximately each 1 min per gas). Also, the atmospheric concentrations are mixed, since the chamber 1 presented higher CO_2 and N_2O concentrations, but lower CH_4. Equally, in chamber 2, the CH_4 and CO_2 concentrations were lower, whilst the N_2O ones were higher.

3.2 Discussion

Given the fact that the data obtained from the sensors is some-how low related with the atmospheric data, it is clear to ensure that these kinds of sensors present bias on their measurements. During the development of the first prototype of the sensor network, we realized that the MQ4 and MQ135 sensors presented an elevated current consumption relative to the total amount provided by the development board. For this reason, the data obtained from these sensors presented, sometimes, values outside the magnitude order of the remaining data.

These data with bias probably caused the non-related results of Table 2 regarding the CO_2 values. The datasheet of the MQ4 sensor indicates a maximum current usage of 250 mA, with a recommended of 150 mA [14]. Therefore, as per the development boards employed 6 analogue inputs, the demand of current should be distributed between all the sensors and, with the relatively high current demand of the gas sensors, these bias values might appear.

On the other hand, the result analysis for the gas samples analyzed in the laboratory when the sensors captured the previous data is presented in Table 3. These results were taken from the work described by Lubo et al. [3].

Table 3. Average sensor values of greenhouse gases inside the chambers analyzed in the laboratory

Chamber 1				
Sampling time	Temperature (°C)	CH$_4$ value (ppm)	N$_2$O value (ppm)	CO$_2$ value (ppm)
7:16 to 7:36	21.6	2.68	0.368	658
7:36 to 7:56	21.7	2.49	0.433	1212
7:56 to 8:16	22.0	2.41	0.330	1686
8:16 to 8:36	22.2	2.42	0.414	1979
Chamber 2				
7:16 to 7:36	21.8	2.66	0.328	642
7:36 to 7:56	22.6	2.48	0.379	1011
7:56 to 8:16	23.2	2.43	0.404	1379
8:16 to 8:36	23.6	2.33	0.455	1718

Comparing the results of the table above with the ones gathered by the sensors (Tables 1 and 2), it is possible to see that the CO_2 concentrations are the more distant between the laboratory and the sensor measurements (e.g., 386 ppm measured by the sensors versus 1979 ppm in the laboratory comparing the highest values in both tables). In general, the measurements of the sensors were considerably lower than the ones analyzed in the laboratory. This implies only one punctual exit: the calibration of the sensors was not the best, since blame the specialized equipment in the laboratory is less probable; nevertheless, this possibility still exists. If an analysis of the gas flux is performed on the data in Table 3, it will also be very different than the one for the sensors data. Consequently, we have summarized some facts that might have affected the sensors measurements (considering the hypothesis that the sensors values were the non-calibrated ones):

- The calibration of the gas sensors was not the appropriate, since they only operated in intermittent periods before the sampling and the 48-h period of preheat was not performed.
- The bias presented in the gas measurements might have affected the global results, since this bias was, occasionally, up to 4 times the real measurement of the sensors.
- This bias was caused by the high amount of current the gas sensors require, forcing the development board to present momentary faults, reflected in these wrong values.

For the following prototypes, we are encouraged to consider the following information towards the achievement of solutions to suppress the previous issues:

- Ensure a correct calibration of the MQ-family sensors, by preheating them 24 h before of the measurements. This calibration can be performed in any environment; it is not necessary to be in the same measurement place.
- To attack the bias problem, we are encouraged to mount a set of transistors to increase the power the development board provides to the sensors. This will be a top priority for the following prototypes.
- With the purpose of increasing the wireless coverage range, we suggest to use either a GSM shield or an RF transceiver with the development board. This will increase in a considerable way the coverage area, since in the assessed sector there is GSM coverage of the main operators in Colombia and operating in a low frequency, the path loss is considerably reduced.

Even though the measurements performed in this project are, somewhat, in accordance with the atmospheric levels, we consider that these results cannot be used to conclude of the adequate performance of the assessed sensors. More tests are required to achieve conclusions about the real performance of the main utility of sensor nodes —like the ones we presented here: open source, cheap, with a large support community, and easy to use— in the case of the Colombian crops.

Considering the results found during the development of this research work, it is possible to affirm that the design, building, and operation of wireless sensor networks to monitor the greenhouse gases concentration is achievable either with low-end devices or with more sophisticated gadgets. The main considerations we can emphasize are related with the calibration of the sensors, with the encapsulation of the elements (by ensuring a correct impermeability), with the selection of the wireless technology depending on the amount of information to deliver, and with the desired variables to measure (the foundation might be this work, but for other crops/activities, substantial changes are probably required.

4   Conclusions and Future Work

The wireless sensor networks are an important field in the enhancement of the agricultural labors. In Colombia, this trend is starting to be applied in several crops and associations, cultivators, and governmental entities are joining efforts to allow the farmers to take advantage of the benefits both the WSN and —in general— the precision agriculture offers. We presented a novel proposal of a network applicable to the sugarcane crops and focused on the measuring of the most relevant greenhouse gases (i.e. CO_2, N_2O, and CH_4). The first results show that the concentrations of these gases are somehow related with the atmospheric concentrations and they diminish as time passes, either inside or outside the closed chambers —used as a reference point—. Even though there was bias in the measurements, caused by the current demand of the gas sensors, the sensor values were, in general, suitable and in accordance with the expected values. Nevertheless, when we performed the comparison between the sensor data and the data analyzed in a laboratory, considerable discrepancies appear.

Differences up to 1600 ppm in the carbon dioxide concentrations and up to 300% in the gases flux calculation indicates that there is a need to calibrate and supply the adequate current amount to these elements.

Regarding the wireless communication, it is clear to say that the selected wireless modules are not very suitable for conditions where the obstacles are dense enough to affect the signal coverage. For our assessed case, the sugarcane had between 2 and 2.5 m high. In this scenario, the coverage area was only within a 75-m radius and in the best of the cases, the maximum radio coverage was an isolated case of 90 m. Consequently, we encourage to use alternative communication devices if considerable distances are required (in this case, 2.26 km).

Our first prototype is an open door to the research in this area, by indicating that the wireless sensor networks build based on open source hardware platforms can be suitable to supply the needs of the Colombian farmers in terms of control, management, and monitoring of the agronomic and climatic variables in a crop. We encourage to consider extra solutions to increase the wireless connectivity for the elements to allow the remote measurements.

The immediate future work entails the upgrading of the system components by acquiring either a GSM shield or an RF transceiver operating at 433 MHz to increase the wireless connectivity. We expect to solve the current issue with a set of transistors on each node to provide the required current for the gas sensors. Besides, the implementation of a web server to store all the data is another pending task to do. Furthermore, we plan to assess other non-closed-source commercial solutions such as the CC1101 device to replace the used radiofrequency elements. That module is even cheaper than the used for this work and it operates in a sub-1 GHz band, which can help in reducing the high absorption problem due to the air humidity (i.e., evapotranspiration) and reduce the losses and increases the coverage.

Furthermore, we are encouraged to continue this research by increasing the energy efficiency of the system by either using more efficient sensors, larger solar panels and batteries, and more efficient wireless transceivers; our goal is to present a research prototype able to be transformed to a commercial solution for farmers.

References

1. Food and Agriculture Organization for the United Nations: FAO Statistical Yearbook 2013: World food and agriculture, Rome (2013)
2. Tey, Y.S., Brindal, M.: Factors influencing the adoption of precision agricultural technologies: a review for policy implications. Precis. Agric. **13**, 713–730 (2012)
3. Lubo, C., Rodríguez, L., Abadía, J., Orozco, O., Sierra, B., Llano, G., López, A.: Sugar cane crop carbon footprint. A contrast between an organic crop, a transition to organic crop and a traditional crop. Presented at the 5th international Ecosummit Congress, Montpellier, France, 29 September 2016
4. Grivet, L., Arruda, P.: Sugarcane genomics: depicting the complex genome of an important tropical crop. Curr. Opin. Plant Biol. **5**, 122–127 (2002)
5. Paturau, J.: Alternative Uses of Sugarcane and its Byproducts in Agroindustries. http://www.fao.org/docrep/003/s8850e/s8850e03.htm

6. Arango, S., Yoshioka, A.M., Gutierrez, V.: Análisis del ambiente competitivo del Cluster Bioindustrial del Azúcar en el valle geográfico del río Cauca: Desarrollo y retos. Multimedios Javeriana Cali, Santiago de Cali (2011)
7. Balaji, S., Murugaiyan, M.: Waterfall vs V-model vs agile: a comparative study on SDLC. Int. J. Inf. Technol. Bus. Management. 2, 26–30 (2012)
8. Zeidman, B.: The universal design methodology–taking hardware from conception through production (2002)
9. Ulgen, O., Black, J., Johnsonbaugh, B., Klunge, R.: Simulation Methodology - A Practitioner's Perspective (2015). http://www.pmcorp.com/Portals/5/_Downloads/SimulationMethodology_APractitioner'sPerspectivev2.pdf
10. Shalloway, A., Trott, J.R.: Design Patterns Explained: A New Perspective on Object-Oriented Design. Pearson Education (2004)
11. The Arduino Company: Arduino Software (IDE), Arduino Environment. https://www.arduino.cc/en/Guide/Environment
12. Vaduva, L.: MQ4 Gas Sensor – Methane Natural Gas Monitor with MQ3/MQ4 Sensors. http://www.geekstips.com/mq4-sensor-natural-gas-methane-arduino/
13. Castro, M.: Understanding a Gas Sensor. http://www.jayconsystems.com/tutorials/gas-sensor-tutorial/
14. Zhengzhou Winsen Electronics Technology Co., Ltd.: Flammable Gas Sensor (MQ-4 Model). https://cdn.sparkfun.com/datasheets/Sensors/Biometric/MQ-4%20Ver1.3%20-%20Manual.pdf

A Proposed Unmanned and Secured Nursery System for Photoperiodic Plants with Automatic Irrigation Facility

Fatima Jannat[1]([✉]), Tasmiah Tamzid Anannya[1]([✉]), Tanu Dewan[1],
Saad Bin Bashar[2], Ayeasha Akhter[1], Ismat Tarik[1], and Farhana Afroz[1]

[1] Department of Computer Science and Engineering, Military Institute of Science
and Technology, Dhaka 1216, Bangladesh
fj.jannat20@gmail.com, anannya.tasmi@gmail.com
[2] Kulliyyah of Information and Communication Technology, International Islamic
University Malaysia, 53100 Selangor, Malaysia

Abstract. Living in a threat of global warming, the effort to make the world a little bit greener is what we all want. Automation of nursery system can make people more encouraged to contribute in making a greener earth. With the advancement of the technologies unmanned or less man controlled nursery system has become very popular. This paper presents an embedded secured smart nursery system that provides automated watering and drainage system, automated light control and supply system for photo-periodic plants and automated light system sensing the light intensity of garden. The design of this system was implemented in an academic environment using soil moisture sensor, LDR sensor and mobile application. The proposed system is easier to set up, cost-effective, fully automatic and has easy maintenance.

Keywords: Automation · Sensor · Mobile application · IoT
Irrigation · Photo-periodism · Smart nursery

1 Introduction

With the advancement of technology, automation is playing a very effective and important role in every sectors. Automation in various sectors is reducing the over load on man power and human error. On the other hand, embedded technologies are drastically changing the definition of automation. In today's world, one of the most concerning issue is global warming and reduction of forestation in an alarming rate [1]. Lack of proper monitoring of plants is one of the main reasons that reduce the growth of plants and thus results in people's disinterest in farming. Smart nursery can reduce these problems with the help of smart data and technologies, which can facilitate and support a sustainable nursery system that help people both living in cities and urban places.

The main part of this research work directs to automatic control of photoperiodic plants and control of irrigation system with security. Photoperiodism is

© Springer Nature Switzerland AG 2019
J. C. Corrales et al. (Eds.): AACC 2018, AISC 893, pp. 136–145, 2019.
https://doi.org/10.1007/978-3-030-04447-3_9

the plant's ability to flower based on the relative day or night length [2]. Considering these conditions plants are divided into short day plants, long day plants and neutral day plants. Though the plants are labeled upon the 'day length', it is found that night length is the dictating factor for flowering a plant. Our research work aims to tackle this complex system in an effective way by creating an automated system, so that, it brings about precise control needed to provide the most proper condition of plant growth.

Irrigation can become a very troublesome task if there is less man power for large farming area. Also the necessity of water varies according to soil dryness. Again without proper watering, the growth of the plant can be hampered in a large scale. So making the system automatic with wireless control can make this task easier. It can lessen man effort and increase people's interest for planting more tree.

Again, in nursery or large farming area, unauthorized persons should not have the access to enter. Plants can get stolen and people can cut down trees for their own profit. So having minimum security in every nursery is always desired. Hence, the goal of this research work is to create an automated, unmanned and secured nursery system which can facilitate the breeding of photoperiodic plants and provides efficient irrigation facilities.

In this paper, the later sections are organized as follows. A brief overview of the related work and literature review is presented in Sect. 2. The overall system architecture of the proposed nursery system is discussed in Sect. 3. In Sect. 4, the overview of total experimental setup and result of the model of the proposed system is depicted, followed by conclusion in Sect. 5.

2   Literature Review

This section reviews few of the previous related various embedded systems where the researchers used various sensors to make their system automatic.

In case of nursery, automated watering system is one of the very demanding features from the users. Manually controlling water system is not always an easy task. In [3], the authors proposed a project where water pump is automatically controlled according to humidity level and temperature. They measured humidity level using humidity sensors. When humidity is more than average value then water pump starts watering. In this paper, when humidity sensors start working then that information is send to the mobile through GSM. The paper [4] also worked with GSM module for a solar automated irrigation system.

The authors in [5] worked on a time threshold and a crop water stress index (CWSI) which is a thermal-based stress index calculated from daylight hours. They have used radiometric sensors and one of the shortcomings of this sensors is that a false positive signal can lead to over irrigation. On the other hand, the paper [6] showed how potential automated irrigation system can be by experimenting with different control strategies. Sap flow readings in the trunk of trees have been used to examine the need of water in paper [7]. The authors have used a CRP consisted with a measurement unit (MU), a control unit (CU) and a pump and electrovalve controller (PEC) for the sap measurement.

The paper [8] showed by using dual crop coefficient approach, an automatic irrigation system can be a good tool to manage water consumption in real time. The authors had implemented a wireless sensor network linked with Granular matrix sensor (GMS). Again in paper [9], color analysis through image processing and computer vision technologies has been used for determining the necessity of crop irrigation.

In [10, 11], the authors emphasized on security according to users need. Users need to enter correct password to open the door and door will open for a specific amount of time. One needs to enter in between this time and one has some time to enter his/her password. When any user enter wrong password then in LCD display equivalent message will be displayed. This micro-controller based password protected locking security system is used in various aspects like bank lockers, home, BIOS locking system etc.

In [12], an advanced light control system has been proposed which is equipped with a photo sensitive detector (LDR-Light Dependent Resistor). This light control system is working based on the amount of light intensity in the environment at that time. According to light intensity using these LDR sensors light is controlled automatically.

In sum, many automation works have been done using sensors to ease the workload of many sectors. This literature study also shows that these different research works have different fields of application where they fit most. But none of these paper worked with the photoperiodic plants which is one of the main focuses of this paper. The proposed work in this paper thus focus on a total model for a embedded nursery system that will have many automatic features integrated in it and also to implement it in an experimental environment.

3   Conceptual Design of the System

A system named *Smart Nursery* was developed to attain the objective of this research. The objective of this work was to develop a full *secured* system that will need minimum (almost zero) number of workman in nursery. Figure 1 shows the diagram of the total experimental setup of the system. The diagram depicts how all the major components including the necessary sensors are connected with MCU (micro-controller unit) and how each output action is triggered in response to any input.

The whole system developed in the smart nursery can be divided into three major parts: (i) Automated Control for Photo-periodic Plants, (ii) Automatic Irrigation System and (iii) Password Protected Security System. The whole system is also integrated with a mobile application.

The features of the whole proposed system are described below.

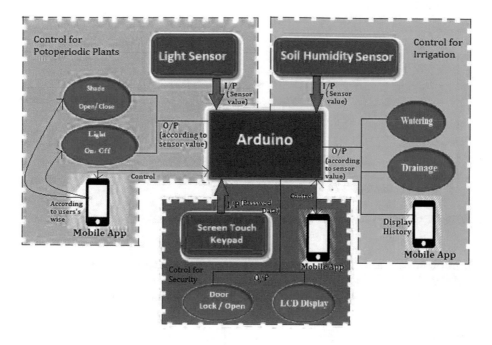

Fig. 1. Block diagram of the total experimental setup

3.1 Automated Control for Photo-Periodic Plants

The control for photo-periodic plants focuses on mainly two cases: short day plant and long day plant. While taking care of plants, people want seasonal or off seasonal flowering to their garden to enhance the growth of their plants. Therefore, this work supports the plants artificially to nourish its proper growth. To go over the flowering state plants requires certain amount of darkness (Fig. 2). But as the day or night period dependent over the earth surface and many seasonal changes, sometimes it becomes hard to achieve that particular amount of dark duration. If long day plants fall over in short day lengths due to natural causes, it is needed to ensure the minimal dark duration artificially by providing the external light source.

This work organizes the proposed system considering two things: intensity of external light source inside the green house and the sunlight. Figure 2 shows the working procedure of the automatic control system for photoperiodic plants. Initially for long and short day plants the system set the start time to zero and critical time as individually requires. For a long day plant if the current start time is greater or equal to the critical time and the intensity of the light source is smaller or equal to 5 (a threshold digital value which is converted from voltage by arduino), then the addition lights turn on or intensity of the source increased. Here, intensity 5 is considered as threshold value of the light sensor as this work is developed in an academic environment, which will vary in a real environment

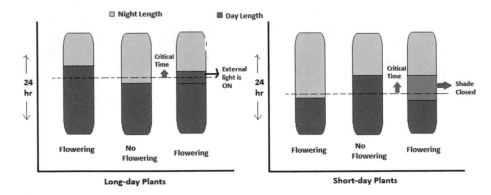

Fig. 2. Graph on impact of external light and shade in case of flowering of plants

from plant to plant. After that, start time is set equal to the current time. The additional LED inside the house lighten or increase their intensity automatically as an external source to ensure the minimum number of lighting hour.

3.2   Automated Irrigation

Soil moisture is a primary factor for healthy growth of plants. Too little moisture can result in plan death whereas too much moisture during rain/flood, also causes root diseases. The required level of soil moisturizing varies for different plants/crops. Considering these reasons, this project focuses on the automated irrigation facility. This work considers two prime cases: 1. Watering while the soil gets dry and plants need water 2. Draining water in case over water level or flooding. At first the level of soil moisture is measured. If the moisture level gets lower than threshold (threshold value will vary depending on the plant), that means if the dryness of the soil increases, a pump starts watering automatically. On the other hand, if the level of moisture gets too much higher than the threshold, it indicates that the water level is also high. Thus, in that case, a pump starts to drain out extra amount of water from land The whole process of this feature is shown in Fig. 3.

3.3   Password Protected Security System

Considering the security of nursery and preventing the illegal cutting down of plants, this proposed system offers a password protected system. Correct password needs to be provided to open the door while entering the nursery or plant area. The password can be entered from door-attached keypad or from the mobile application. The system will generate a warning to the owner if wrong password is given two times, in case any intruder tries to break in.

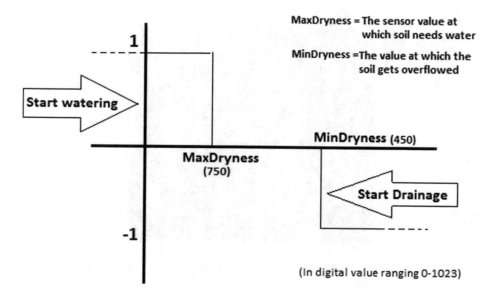

Fig. 3. Starting of watering and drainage systemlabelfig

3.4   Control Through Mobile Application

The whole system can be controlled via a developed mobile application for remote access. The condition of nursery can be monitored via the app and the app also allows the users to remotely access the system. The start and stop of irrigation can also be handled automatically using mobile application. When an user click the "START WATERING" button in the mobile application, the pump starts watering the soil from a water source. Again when "START DRAINAGE" button is clicked, another pump starts to drain water from the over-flooded area. History of start and end of irrigation can also be seen from the mobile application.

4   Experimental Setup and Result

This section gives a broad view of the implemented model of the proposed system. A prototype version of this system was primarily developed and tested in an academic environment in context of Bangladesh. For the development of the entire system two sequential steps have been used. (1) Development of the hardware, (2) Development of the mobile application. The common parameters for hardware implementation considered are temperature, humidity, light intensity, moisture and pH value of soil. The mobile application for remote controlling has been developed in MIT App Inventor 2. This is an open source cloud-based tool for developing mobile application for the Android operating system. The application has different user interfaces (UI) for different control parts. The app communicates with the micro-controller unit (MCU) using a HC-05 Serial Port

Fig. 4. Image of the developed prototype with turned on LED light compensating the required light for flowering

Bluetooth module. In the model, two houses were build where one house is for long-day plants and other is for short-day plants. The house (Fig. 4) which is for long-day plants is built up with glass. And the house for short-day plants is a closed house with a opened hood at the top. In case of photoperiodism, there is a critical day length and for short-day plants the day length has to be smaller than the critical time whereas for long-day plants day length have to be longer. Thus at first we count the start time of flowering, and then compare it with the critical time. With the counting time, we also measure the light intensity. For long-day plants if the current time exceeds the critical time and if there is darkness, we will need more light for flowering. After exceeding the critical time if the darkness is remained, then the light inside of the house will be become ON. Otherwise light will be remained OFF. For short-day plants if the current time equals or greater than the critical time, and also there is light outside, then we need to provide darkness for flowering properly. So when the current time becomes equal to the critical time, the opened hood will be closed for providing darkness.

In case of handling automated lighting system, we have used LDR Photo Resistor that sense the light intensity. Input of this light intensity passes through the MCU then it decides whether it requires additional light or not. Additionally there is another feature by which one can also control the light by the mobile application. The model of the prototype system for photoperiodism is given in Fig. 4.

For automated watering and drainage system, we have used Soil Moisture Sensor (SEN-0018) which has two exposed pad that actually sense the soil moisture. Soil Moisture Sensor send measuring data to the MCU. According to those data, the MCU decides whether to start watering or draining. This data history

History

Garden Watering History

Connect

Connected

Soil Moisturiser : 1017 – Dry … Start watering
977 – Dry … Start watering

Fig. 5. Mobile App history of irrigation system

transfer to the mobile application by which one can easily track the up-to-date data (Fig. 5). The watering and drainage system works through Micro-pump motor(PM3712). Also, from the mobile application one can also start and stop irrigation by clicking on app buttons.

For making the nursery more secured, the control of security system has been made password protected using a $4 * 4$ matrix 16 keypad module (Fig. 6), LCD-1602 with Blue Backlight and Micro Metal Gear Servo. After entering password in keypad data passes to the MCU and it decides whether it matches with the password in the database or not. Given input can be seen on the LCD. If the enter password is matched then the door automatically opens and closes after a while which is controlled by a servo motor. The willing person must enter within this particular time. Besides, one can enter password through mobile application to enter the nursery. Figure 6 shows the prototype developed to implement the password protection for the proposed nursery system.

Fig. 6. Password protected secured system

Table 1. Required equipment list for the proposed system

Equipment	Quantity
RF-370CA-22170 Micro-pump Motor (PM3712)	2
Arduino Uno R3	4
Soil moisture Sensor	1
LED (Red, Green)	2
LDR Photo Resistor	2
4*4 Matrix 16 Keypad Module	1
HC-05 Serial Port Bluetooth	2
Micro Metal Gear Servo	2
LCD-1602 with Blue Backlight	2

Table 1 shows the equipment list that is needed while implementing the proposed system.

The result of the proposed system is evaluated by setting up all the equipment for a certain time and by observing all the output actions with respect to changing variables. The supply of extra light and shade in case of photo-periodic plants vary according to the existing amount of light and the timing of the light. In case of irrigation, the performance is smooth and swift. The water flow is controlled by setting up some delay. Thus, a large amount of water do not get flowed at once. The experimental model successfully provides an automatic and complete unmanned nursery system.

5   Discussion and Conclusion

This paper proposes a full automatic system for smart farming having features for photo-periodic plants with irrigation facility and nursery security. One of the prime goals of this paper is to minimize the man power in farming through automation. The control system for photo-periodic plant and the control system for automatic irrigation, will undoubtedly increase the growth of plants. To the best of our knowledge, no work has considered photo-periodic plants in this intelligent way yet. Through this proposed system, photo-periodic plants can be grown throughout the whole year as this system provides right amount of light required for specific plants. Security in nursery is also an important feature specially for the people who grow plants commercially. Again to prevent deforestation and stealing trees or plants security is highly appreciating. The mobile application that has been developed in this proposed system, will increase user's monitoring by providing record and data.

The proposed research work has a few limitation as well: the system was developed as a prototype system in an academic environment and no study focusing on user (owner) experience was conducted. The results of the evaluation study may vary depending on the plants, environment and the time to conduct the experiment. Future work would be conducted so that the prototype system can be converted into a concrete system to conduct a few more experiments in a real life with different type of plants in the different times. Moreover, a usability study would be conducted to in the future to assess the owners experience and intention to adopt such system. This sort of smart farming will increase the forestation and people's interest to grow more plant as it makes the farming more easier.

References

1. Scott, C.E., Monks, S.A., Spracklen, D.V., Arnold, S.R., Forster, P.M., Rap, A., ijl, M., Artaxo, P., Carslaw, K.S., Chipperfield, M.P., Ehn, M., Gilardoni, S., Heikkinen, L., Kulmala, M., Petj, T., Reddington, C.L.S., Rizzo, L.V., Swietlicki, E., Vignati, E., Wilson, C.: Impact on short-lived climate forcers increases projected warming due to deforestation. Nat. Commun. **9**, Article number 157 (2018)
2. Garner, W.W., Allard, H.A.: Further studies in photoperiodism, the response of the plant to relative length of day and night. J. Agric. Res. **23**(11), 871–920 (1923) ISSN 0095-9758
3. Modi, D.: Automatic Intelligent Plant Irrigation System using Arduino and GSM board, September 2015
4. Rehman, A.U., Asif, R.M., Tariq, R., Javed, A.: GSM based solar automatic irrigation system using moisture, temperature and humidity sensors. In: International Conference on Engineering Technology and Technopreneurship (ICE2T), Kuala Lumpur, pp. 1–4 (2017)
5. O'Shaughnessy, S.A., Evett, S.R., Colaizzi, P.D., Howell, T.A.: A crop water stress index and time threshold for automatic irrigation scheduling of grain sorghum. Agric. Water Manag. **107** (2012) https://doi.org/10.1016/j.agwat.2012.01.018
6. Romeroa, R., MurielbI, J.L., García, I., Muñoz de la Peñac, D.: Research on automatic irrigation control: state of the art and recent results. Agric. Water Manag. **114**, 59–66 (2012)
7. Fernández, J.E., Romero, R., Montaño, J.C., Diaz-Espejo, A., Muriel, J.L., Cuevas, M.V., Moreno, F., Girón, I.F., Palomo, M.J.: Design and testing of an automatic irrigation controller for fruit tree orchards, based on sap flow measurements. Aust. J. Agric. Res. **59**, 589–598 (2008)
8. Cancela, J.J., Fandiño, M., Rey, B.J., Martínez, E.M.: Automatic irrigation system based on dual crop coefficient, soil and plant water status for Vitis vinifera (cv Godello and cv Mencía). Agric. Water Manag. **151**(C), 52–63 (2015)
9. García-Mateos, G., Hernández-Hernández, J.L., Escarabajal-Henarejos, D., Jaén-Terrones, S., Molina-Martínez, J.M.: Study and comparison of color models for automatic image analysis in irrigation management applications. Agric. Water Manag. **151**(C), 158–166 (2015)
10. Sriharsha, B.S., Vishnu, S.B., Sanju, V.: Password protected locking system using Arduino. BVICAM's Int. J. Inf. Technol. **8**(1), 959–964 (2016)
11. Annapurna, L.B., Mounika, K., Chary, K.C., Afroz, R.: Smart security system using Arduino and wireless communication. Int. J. Eng. Innov. Res. **4**(2) (2015) ISSN 2277 5668
12. Rath, D.K.: Arduino based smart light control system. Int. J. Eng. Res. Gen. Sci. **4**(2) (2016) ISSN 2091-2730

Design and Development of an Intelligent Seed Germination System Based on IoT

Muhammad Nazrul Islam[✉], Mahmuda Rawnak Jahan, Abid Ali,
Shamsuzzaman Rony, Tasmiah Tamzid Anannya, Faisal Ibn Aziz,
Moin Bayzed, Anika Yeazdani, and Md. Fazle Rabbi

Department of Computer Science and Engineering, Military Institute of Science
and Technology, Mirpur Cantonment, 1216 Dhaka, Bangladesh
nazrulturku@gmail.com, nitu.islam96@gmail.com

Abstract. Agriculture is heavily dependent on the weather and soil condition. Day by day the weather is getting unpredictable due to climate change, thus the soil condition gets changed. Thus the use of Information and Communication Technology (ICT) has become a key component in agriculture to improve productivity. The objective of this research is to develop an Internet of Things (IoT) system, named Smart Germination Assistant (SGA) system to improve the seed germination for different types of plants. The proposed system uses multiple sensors to measure and automatically adjust the moisture, humidity, pH level, temperature, and sunlight of a land to the standard value for each particular type of crop. The performance of the SGA system was experimentally evaluated with 300 seeds of jute leaf, where 150 seeds were sowed in SGA system and other 150 seeds were sowed in a natural environment. The experimental data was collected during 25 May to 07 June, 2018. The experimental results showed that the SGA system makes seed germination at higher rate with maximum survivability.

Keywords: Agriculture · Climate change · Natural germination
Internet of Things · Intelligent system · Sensors

1 Introduction

The emergence of a new plant from a seed is known as germination. For plants that are reproduced through seeds are eventually grown into young plants through the process of seed germination. While the seeds are being planted, they remain inactive until conditions are suitable for germination. For germination to occur, various conditions must be met such as the proper amount of water, oxygen, moisture, temperature, and light. When these conditions are met, the seeds begin to enlarge as it takes in water and oxygen from the environment. Then the seeds' coat break open and roots emerge from the seeds, which is followed by a plant shoot. This initial stage of a plants development is germination. Now not always the desired environment for plant can't be achieved. This is when IoT

© Springer Nature Switzerland AG 2019
J. C. Corrales et al. (Eds.): AACC 2018, AISC 893, pp. 146–161, 2019.
https://doi.org/10.1007/978-3-030-04447-3_10

comes into play. With the help of Iot a favorable condition for seed germination can be achieved. Whenever the environment gets inclimate, there has to be a system that is able to change the parameters of the environment suitable for seed germination.

During the last decades, farmers' understanding and observation in the process of seed germination has been expanding. In [1], author Jaime Kigel and Gad Galili have shown the recent modern approaches related to genetics and molecular biology showed significant improvement in seed research which includes germination of seed. In [2], State of the art technologies such as automatic exhaust fan, high tech air conditioning system, ambient lighting apparatus, smart plant watering machine and etc. are being used to aid seed germination process. In context of Bangladesh those are proved to be expensive and inefficient. Therefore, under developed countries like in Bangladesh, the farmers are often either not known to the modern technologies or they can't afford. Agriculture in this country mostly depends on weather which puts a great effect to growth of plants, specially during the germination period which is the most delicate period of plantation. Due to global warming the climate behaviour has changed. Unexpected rain and storm are interrupting the germination and plantation process quite heavily. On the other hand, when the dry spell continues for too long time then droughts happen. At that time seed germination becomes impossible. Automated system is needed to water them, as well as for maintaining all other attributes for a successful germination. There are only a few Agro-farms growing plants in customized environment on a small scale of land. Implementing their system to a landmass will cause huge expenditure and raises a big risk issue in case of failure. One of the leading Agro-farm companies in Bangladesh named "Keya Group" [3] that is taking care of only vegetables starting from its germination to plantation in a very expensive and ineffective way.

The Internet of Things (IoT) is a recent innovation that connect any physical devices with the computer systems to enable these devices as well as to exchange data [4]. This technology provides a network between the physical world and the computer-based systems, which in turn improve efficiency and cost-effectiveness. Thus, the IoT technology can provide such favorable condition for seed germination to make the process automatic, profitable and effort-less. In such a case, few basic components and sensors can be connected to create the favorable condition for better seed germination with an increased survival rate of seeds.

Therefore, the objective of this research is to develop an intelligent system based on Internet of Things (IoT). A number of sensors are used to detect all the rudimentary parameters to aid seed germination. The system is designed to alter the values in order to get the environment which is worthy to the seeds to germinate. In other words the aim of the research is to maximize seed germination for different types of plants in a more cost effective and efficient manner in a landmass.

The remaining sections of this paper are organized as follows. The existing work focusing to automated germination systems are presented in Sect. 2. The conceptual design of the proposed system followed by how the system was implemented is discussed in Sect. 3. Section 4 presents study procedure, data analysis and findings of evaluation study. The discussion and concluding remarks are presented in Sect. 5.

2   Related Works

Although a number of research have been carried out focusing to automatic plant maintenance, only a few studies explicitly focused on seed germination. Germination is as important as plant maintenance and growth. According to [5], the conditions that must be present in order to properly germinate a seed includes proper temperature, moisture, oxygen, and darkness. There is also great effect of acidity on plant germination and growth which is known as pH in [6].

A study in [7] was carried out on 30 different vegetables suggests to put a heat mat for robust and faster growth of the seeds. In [8,9], the heat mat was not efficient due to its cost and lack of availability. In [10], a chamber was used to preserve all the seeds and germinate where the temperature was manipulated from the outside. But the chamber system was unable to control other parameters like providing sunlight, moisture and humidity.

In [11], a saucer tray filled with water was put under the plants so that it always remains moist during dry season. This system was successfully simulated on plotted plants only. Mulch is used in [12] to control moisture in soil and experimented by distributing considerable amount of mulches around the plants. But it has been found that mulch retards the heating of the soil by the sun. It does not create problem but in early or mid-spring, when its need the warmth (that can get from the sun's rays) it can inhibit the germination of seeds, especially for those seeds what need a higher soil temperature for germination. Lower temperature of soil reduces the growth rate of seedlings. Mulching does give a great cover for small slugs, which can be devastating for little seedlings. Also at times it can be unsuitable for crops that need fine sandy soil to flourish.

Other researches used closed boxes containing saturated solutions of certain salts as simple methods of providing atmospheres of known humidity's so that materials and components enclosed in the boxes may be given a standard condition of moisture content. But a series of experiments have shown that the method cannot be relied upon to produce atmospheres of known humidity's except for loads which absorb moisture at a negligible rate. Even a rate of moisture absorption as low as $4 \, mg \, h^{-1}$ is sufficient to depress the humidity by about 2% can be observed in different studies [13,14].

The effects of light quality on lettuce leaf and spinach were examined in [15]. The plants were grown in an environmentally controlled room with a 12-h light period and under a light/dark temperature of 201 C/181 C and a photosynthetic photon flux density (PPFD) of $300 \, mol \, m^{-2} s^{-1}$ under four light quality treatments, i.e., red light from red fluorescent lamps (R) or blue light from blue fluorescent lamps (B) or a mixture of R and B (RB) or white light from white fluorescent lamps (W). These experiments show that controlling light quality is useful to achieve higher productivity or higher nutritional quality of the commercial crops. Another study in [16] investigated the effect of 100, 50 and 25%

Table 1. An overview of existing works

Name	Objective	Technology	Limitations
Compact germination Station [7–9]	Moist and heat able environment	A waterproof mat, humidity dome and a watertight based tray	Lights cannot be controlled
Heat mat [10]	Maintain optimum temperature	A heat mat has been fitted under the soil	Heat exhaustion is not considered
A saucer tray [11]	Continuous water flow for soil moisturizing	The tray is refilled with water after a period of time	Can be used for potted plants only, has to be refilled manually
Control of atmospheres by saturated salt solutions [13,14]	Provides atmosphere of standard humidity	Sealed chambers containing saturated solutions of different salts	Reliable only for seeds which absorb moisture at a negligible rate
Effect of light quality [15]	Tests different light qualities on seeds	An environmentally controlled room with a 12-h light period temperature of 201 C/181 C And a PPFD of 300 mol m^{-2}s^{-1} with four light quality treatments	Effectiveness of light quality treatment differs for different plant species
Effects of nutrient addition and acidification [17–19]	Shows the effect of PH value	Changes nutrient elements and availability during growth	High PH negatively affected most of the germination

light intensities on seed germination with a particular humidity and temperature and the seedlings have almost highest growth value among the species with their required light intensity.

A study conducted in [17] showed that, the complex ecological effects arising from sulphur, nitrogen and phosphorus deposition that change species richness and community composition. Another study in [18] from US National Library of Medicine assessed the separate and combined effects of nitrogenous compounds and pH on the percentage and rate of germination of seeds of different seeds and found high pH negatively affected the germination rate of seeds from most species, but had no effect on the per cent germination of any of the species. Again, an experiment conducted in [19] has showed that, pH is a very important factor in the germination process of beans. A summary of some important related work is given in Table 1.

In sum, the existing systems have its own merits and demerits to assist the germination process. One issue is clearly visible that not all the required parameters like temperature, moisture, oxygen, soil pH and darkness were considered in a single system to make an automated intelligent system for helping germination process. Thus, this research is focusing to develop such an automated system to assist the germination process to improve the seed germination for different types of plants.

3   Design and Development of the System

A system named "Smart seed Germination Assistant (SGA)" was developed to attain the objective of this research. A prototype version of this system was preliminary developed and tested in context of Bangladesh. The common parameters considered for seed germination are temperature, humidity, light intensity, moisture and pH value of soil. The proposed system includes a database that stored the name of different seeds and corresponding ideal values to each of the common parameters. Again, the proposed system will continuously measure the values of these parameters from current environment using specific sensors. Then these collected data was compared with the ideal values of a specific seed as stored in the database and takes necessary actions to maintain the values of germination parameters. The system used (a) two fan for controlling the temperature, (b) a light for controlling light intensity and temperature, (c) a water pump for water supply to increase the soil moisture, (d) a humidifier for controlling humidity. The parameters' values will update continuously through wireless sensors and show in the display device placed at user end. The sensors real-time values will be displayed in the display device. The intended seed for germinating can be selected by user from stored options in the proposed system. Figure 1 represents a flow chart of the proposed system. The key features of the proposed system and it's development are discussed in the following sub-sections.

3.1   Features of the System

The features of the proposed SGA system are given below:

  i. *Selection of Seed:* Different seeds and their required growing environment are stored in the database in a numbered way. To initialize the system at the beginning, the intended seed type is selected by user from the list shown in the display.
 ii. *Temperature Sensing and Control:* The room temperature is being continuously sensed by a temperature sensor. Comparing it to the ideal values the system controls the temperature through an exhaust fan (for decreasing temperature) and a bulb (for increasing temperature).
iii. *Light Intensity Measurement and Control:* The light intensity is continuously measured by a Light Sensor and the intensity is maintained as required by turning a bulb on and off.
 iv. *Humidity Sensing and Control:* The humidity is also continuously sensed by a Humidity sensor and compared with the ideal value. And to maintain the ideal humidity, we have used a humidifier.
  v. *Soil Moisture Measurement and Control:* The soil moisture is measured by a Moisture sensor continuously at a certain time interval. And to maintain the ideal soil moisture for seed the system used an automated water pump to supply water to the soil to increase moisture.

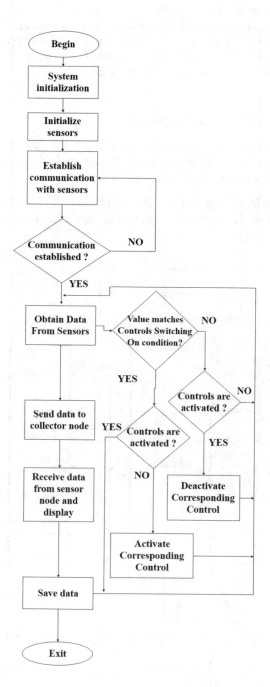

Fig. 1. A flow chart of the proposed SGA system

vi. *Soil pH Measurement and Recommendation:* The soil pH can be measured with a given pH sensor and comparing with the ideal pH, our system will recommend to use appropriate fertilizer taking stored data from database.
vii. *Data in Display:* The system can display the instantaneous values taken from different sensors by a Display device.

3.2   Implementation of the System

The implementation of the system included both the hardware and software. Figure 2 represents the block diagram of the proposed SGA system. The system consists of different hardware devices, that includes:

i. *Temperature-Humidity Sensor:* Temperature and Humidity are both measured in this sensor. System continuously compares ideal data with the ones measured from sensors.

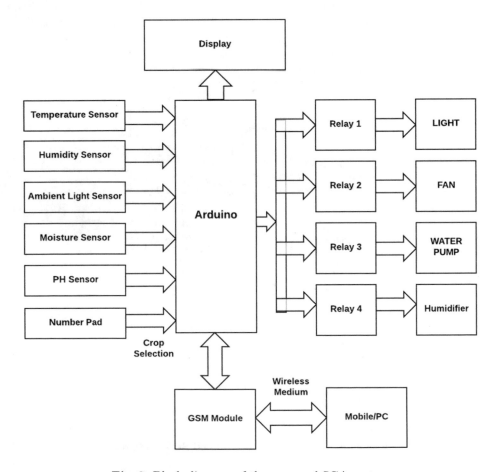

Fig. 2. Block diagram of the proposed SGA system

ii. *Ambient Light Sensor:* This sensor senses intensity of light, following the conditions, there is a bulb connected to the relay that goes on and off.

iii. *Moisture Sensor:* Moisture Sensor is also set in the system with micro-controller that continuously senses the moisture of the soil.

iv. *PH sensor:* The pH sensor is implemented with the micro-controller and it shows the value of pH in soil.

v. *Liquid Pump:* A liquid pump was used having a hose pipe connected to it with a ring, the pump is connected to the relay. This allows the pump to activate only when it's necessary.

vi. *Humidifier:* This is an output device implemented into the system which activates under obligatory circumstances. A relay is connected to the device which works as a switch and switch on the humidifier when percentage humidity goes lower than ideal range.

vii. *Bulb:* To control the light intensity a tungsten bulb is used as a replacement of light and to increase temperature in the experiment box.

viii. *Exhaust Fan:* For controlling the temperature and humidity two exhaust fan is used in the system.

ix. *Display:* A display device shows the data read from different sensors. It has connection with the micro-controller that sends all data from sensors to the device.

x. *Relay Module:* Different output devices; e.g. Fan, Bulb, Motors are given connection with the relay which works as a switch.

xi. *Micro-controller:* The micro-controller is the bridge between software and hardware. It's pins are connected to different sensors and then the data from the sensors are sent to the microprocessor of it which then regulates the controlling devices.

xii. *GSM Module:* The GSM module is connected to the micro-controller to store and retrieve data from the database in a wireless manner.

xiii. *Experiment Box:* All the hardware elements are integrated in an experiment box made of glass and wooden roof with scope of proper ventilation.

The necessary programming code was implemented in micro-controller platform to read the data from sensors. The software system compares the sensors data with the threshold data stored in the system database to instruct the output devices to run. The GSM technology was used to send and retrieve data from database through wireless network. The real time values of different sensors are automatically saved in database and compared with the ideal value stored in database. Figure 3 represents the implemented prototype of the SGA system.

Fig. 3. Prototype of the proposed Smart Germination Assistant (SGA)

4   Evaluating the System

An experiment was conducted in an academic outdoor environment to evaluate
and compare with the system's performance. The proposed Smart Germination Assistant (SGA) system is a glass house with controlled environment with
a provision of passing natural light and air. A total of 300 seeds of Jute leaf
(local name: 'Patshak' (in Bangladesh), scientific name: Corchorus Capsularies,
and family name: Tiliaceae) were sowed at the same time both in the proposed
Smart Germination Assistant (SGA) system and in a natural environment. The
experiment was carried out in Dhaka, the capital of Bangladesh with an elevation
of 22 m height. Average temperature was around 31 °C since it was in the summer. The experimented seeds were under direct sunlight without any intensive
care. The prime objective was to see the difference of the experiment with the
SGA, to see if the SGA gives the better output with intensive care or not. The
SGA was also set in a similar condition, but the only difference is the conditions
were manipulated and set to the suitable values for seeds to germinate fast. To
maintain the confound variable the same type of soil used in both conditions
and equal number of randomly choose seeds (150 seeds/condition) were sowed
in both conditions. The seeds were sowed on 25 May 2018 and the data related
to number of seeds germinated, maximum height, dying rate, and soil condition were collected on different day during the first two weeks (from 25 May
2018 to 7 June 2018). The study data for these two weeks are showed in Table 2
and the difference between two systems on a date of first week (27 May 2018)

Table 2. Experimental data for both natural system and SGA of germination

Date	Day	Data type	Natural system	SGA system	Figure no
25 May 2018	1	Amount of seed sowed	75 gm (150 seeds approximately)	75 gm (150 seeds approximately)	
27 May 2018	3	State of seeds	Yet to germinated	Germination Occurred	Figure 4(a), (b)
29 May 2018	5	No of germinated seed	29	53	Figure 7(a), (b) (See Appendix)
30 May 2018	6	No of germinated seed	68	98	Figure 8(a), (b) (See Appendix)
01 June 2018	8	Soil condition	Drying (pH 4, soil moisture 30–40%), need to be watered	Wet and perfect (pH 4–4.5, soil moisture 65–70%)	Figure 8(c), (d) (See Appendix)
02 June 2018	9	Height of seedling	1.5 cm	3.755 cm	Figure 8(e), (f) (See Appendix)
03 June 2018	10	No of seedlings	75	103	Figure 9(a), (b) (See Appendix)
04 June 2018	11	Height of seedlings	3 cm	5 cm	
05 June 2018	12	No of seedlings	30	95	
06 June 2018	13	Soil condition	Drying (soil moisture 20–25%)	Perfect (pH 4–4.5, soil moisture 65–70%)	
07 June 2018	14	Height of seedlings	5.5 cm	7.6 cm	Figure 4(a), (b)

and a date of last week (7 June 2018) are shown in Fig. 4. The progress of the whole seed germination for both natural system and SGA on different dates within these two weeks are presented in Appendix A.

The results showed that number of germinated seeds were consistently larger in case of SGA system than the natural system. It is clearly visible from Table 2 that within the first 9 days, about 70% of total seeds were germinated in case of SGA system, while only 50% seeds were germinated successfully in case of the natural system. Again, by the 14th day dying rate of seedlings was noticeable higher in case of natural setting (SGA dying rate 17.5% and natural dying rate 40%). The resultant data from Table 2 showed that the success rate of seed germination for SGA and Natural system were 63.33% and 20%, respectively. These results indicate that the success rate of SGA system was more than three times comparing to the natural system whereas the dying rate was noticeably

(a) Natural Germination (27 May, 2018)   (b) Smart Germination (27 May, 2018)

(c) Natural Germination(7 June, 2018)   (d) Smart Germination (7 June, 2018)

Fig. 4. Photos of SGA system and natural germination system on different dates

lesser than the natural system. The soil and leaf conditions observed in different days also showed better in case of SGA system. The Fig. 5 clearly shows that the SGA curve clearly dominates the Natural system in terms of number of seeds that are being germinated in two weeks. The Fig. 6 shows the height of the germinated seeds in two weeks where the SGA lines are always staying ahead of the natural system lines.

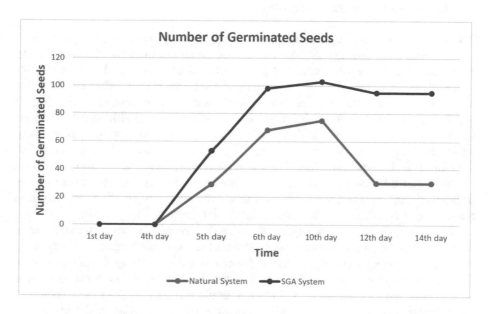

Fig. 5. Number of germinated seeds in SGA and natural system

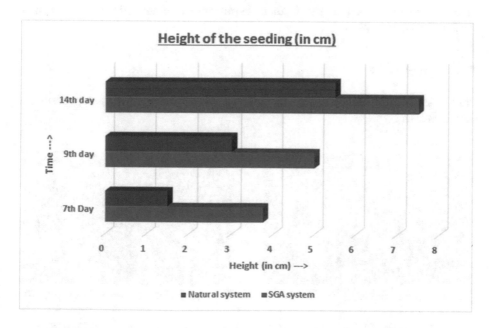

Fig. 6. Maximum height of the seedlings for both SGA and natural system

5   Discussion and Conclusions

The research proposed and developed a smart germination assistant (SGA) system for the seed germination in context of Bangladesh. The real-world data collected from the experimental study showed that the developed system is effective and efficient than the natural system of seed germination, that includes, for example, increases the success rate in seed germination process, reduction of the dying rate of seedlings, increase the growth of the seedlings, less amount of human assistant is required, reducing the human assistant and monitoring effort. In Bangladesh, the weather condition is continuously deteriorating that may bring a devastating effect (high temperatures, localized flooding, etc.) on the seed germination process. Moreover, Bangladesh is one of the most densely populated countries in the world and the area is also very small. Therefore, this research will contribute to the agricultural extension in Bangladesh.

The research has a few limitation as well: the system was developed as a prototype system; only one evaluation experiment was conducted; only one kind of seed (jute seed) was used in the experiment; and no study focusing on user (farmer) experience was conducted. The results of the evaluation study may vary depending on the seed type, location, environment and the time to conduct the experiment. Future work would be conducted to materialize the prototype system into a concrete system and conducted a few more experiment in the different times, location and with other types of seed to assess the effectiveness and efficiency's in a broader scale. Moreover, a usability study would be conducted to in the future to assess the farmers experience and intention to adopt such system for germination.

A   Appendix

Photos taken of different data of the SGA and natural germination system are shown below.

(a) Natural Germination (29 May, 2018)   (b) Smart Germination (29 May, 2018)

Fig. 7. Photos of SGA system and natural germination system on 29 May 2018

(a) Natural Germination (30 May, 2018)   (b) Smart Germination (30 May, 2018)

(c) Natural Germination (1 June, 2018)   (d) Smart Germination(1 June, 2018)

(e) Natural Germination(2 June, 2018)   (f) Smart Germination(2 June, 2018)

Fig. 8. Photos of SGA system and natural germination system on 30 May, 01 June and 02 June 2018

(a) Natural Germination(3 June, 2018)   (b) Smart Germination(3 June, 2018)

Fig. 9. Photos of SGA system and natural germination system on 03 June 2018

References

1. Kigel, J., Galili, G.: Seed Development and Germination. M. Dekker, New York (1995)
2. Seedman.com. https://www.seedman.com/seedstart.htm. Accessed 29 May 2018
3. Keya Agro Process Ltd. Nursery. http://www.keyagroupbd.com/keya-agro-process-ltd/. Accessed 05 June 2018
4. Vermesan, O., Friess, P. (eds.): Internet of Things: Converging Technologies for Smart Environments and Integrated Ecosystems. River Publishers, Aalborg (2013)
5. Seed Germination Method. http://www.fullbloomhydroponics.net/how-to-germinate-seeds/#ixzz5ByerchAR/. Accessed 05 June 2018
6. Acidity Effect on Growth & Germination. http://www.saps.org.uk/saps-associates/browse-q-and-a/592-how-does-acidity-affect-plant-growth-and-germination. Accessed 07 June 2018
7. Grow Great Vegetables. http://www.growgreatvegetables.com/plantinggrowing/germination/. Accessed 07 June 2018
8. Gardener's Supply Company. https://www.gardeners.com/how-to/when-is-it-warm-enough-to-plant/9029.html. Accessed 15 June 2018
9. Allotment & Gardens. http://www.allotment-garden.org/gardening-information/best-temperatures-for-seed-germination/. Accessed 17 June 2018
10. Seed Germinator. http://www.bionicsscientific.com/test-chambers/seed-germinator.html. Accessed 17 June 2018
11. Savvy Gardening. https://savvygardening.com/tips-keeping-gardens-hydrated-heat-summer/. Accessed 19 June 2018
12. Soil Moisture: How to Deal with Excess and Lack. https://www.diynatural.com/soil-moisture/. Accessed 19 June 2018
13. Martin, S.: The control of conditioning atmospheres by saturated salt solutions. J. Sci. Instrum. **39**(7), 370 (1962)
14. O'Brien, F.E.M.: The control of humidity by saturated salt solutions. J. Sci. Instrum. **25**(3), 73 (1948)
15. Ohashi-Kaneko, K., Takase, M., Kon, N., Fujiwara, K., Kurata, K.: Effect of light quality on growth and vegetable quality in leaf lettuce, spinach and komatsuna. Environ. Control Biol. **45**(3), 189–198 (2007)

16. Aref, I.: The Effects of Light Intensity on Seed Germination and Seedling Growth of Cassia fistula (Linn.). Enterolobium saman (Jacq.) Prain ex King. and Delonix regia (Boj) Raf (2002)
17. Roem, W.J., Klees, H., Berendse, F.: Effects of nutrient addition and acidification on plant species diversity and seed germination in heathland. J. Appl. Ecol. **39**(6), 937–948 (2002)
18. Seed germination in response to chemicals: effect of nitrogen and pH in the media. https://www.ncbi.nlm.nih.gov/pubmed/16850869. Accessed 20 June 2018
19. Shields, B.: The Effects of pH Levels in Water on Bean Germination. https://www.enotes.com/homework-help/how-do-different-ph-levels-affect-germination-337812. Accessed 1 July 2018

A Modeling Infrastructure Based on SWAT for the Assessment of Water and Soil Resources

Pierluigi Cau[1]([⊠]) and Giovanni De Giudici[2]

[1] CRS4, Loc Piscina Manna, Ed. 1, Pula, Italy
plcau@crs4.it
[2] University of Cagliari, Cagliari, Italy
gbgiudic@unica.it

Abstract. Many regions around the world, and the Mediterranean basin is just one example, are characterized by scarcity of water and limited productive soils. Agricultural practices in such areas are being based on the intensive use of fertilizers and on the use of limited water resources. Contamination levels in water bodies, both ground and surface waters in Mediterranean countries, are found to be increasing and local and regional authorities are now facing this awkward environmental problem.

In the frame of the regional TESTARE and of the EU SUPREME funded projects, we aim at supporting the set up of a sustainable agricultural production frame, addressing vulnerable communities living in semi-arid and arid areas within the Mediterranean. The scope is to progress web-based observation systems through the integration of state-of-the-art leading edge characterization, monitoring and modeling tools. The modeling infrastructure is being build and here presented to analyze the water, sediment and nutrient cycle at the catchment's scale, and crop growth. In this work, we particularly address the problem of agricultural drought and the impact of agriculture on water quality in Sardinia (Italy, Europe). Local communities have been increasingly challenged by water scarcity that is lowering agricultural productivity. Different soils and crops are considered in the test site.

Keywords: SWAT · Water cycle · Soil management · TESTARE
Climate change · Hydrology · Alto framework · Agriculture
Earth observation products

1 Introduction

Global demand for agricultural production is expected to double by 2050 [1]. This will shape agricultural markets and production systems. Farmers are challenged to produce more but conversely to lower the use of water for irrigation, fertilizers and pesticide. As stated by the United Nations in Paris 2015, sustainability requires the reduction of agricultural pollution in respect of the ecological status of the environment and human health. The increase of food production needs to be harmonized with natural resources and habitats, by limiting the negative effect of the use of chemicals, and improving

© Springer Nature Switzerland AG 2019
J. C. Corrales et al. (Eds.): AACC 2018, AISC 893, pp. 162–176, 2019.
https://doi.org/10.1007/978-3-030-04447-3_11

water quality. Expected climate change scenarios for the Mediterranean show less water availability, prolonged drought periods, heat waves and more intense precipitation events. Uncontrolled pesticide applications, the decline of vital pollinator populations as recently stated by the International Commission on Plant Pollinator Relations (ICCPR - https://www.icppr.com/), and a drier climate are the drivers impacting future food security.

In many EU countries, soils that were considered productive in the past are now considered marginal lands that are of little agricultural value because of by one side adverse climate conditions (e.g. drier climate) or poor soil quality and on the other side because of poor water supply, pollution from previous human activities, or excessive distance from means of transportation. Intensive agriculture in Mediterranean regions is resulting in soil impoverishment, and, water, mostly groundwater, over exploitation. Shallow aquifers are being depleted and sea water intrusion has been detected in many sedimentary aquifers close to the sea.

Cyprus for example offers a paradigm of fertilizer peruse. In fact, despite the fact that a decrease on N fertilizers was there achieved since 1992, the average Gross Nitrogen Surplus per ha in the period 2009–2012 was the highest among the EU-28 member states and accounted to 195 kg N/ha, while for other EU countries it falls in the range 27–120 kg N/ha. In South Mediterranean areas available data on NPK fertilizer consumption are greatly variable, indicatively Tunisia consumes 41, Algeria 11 kg NPK/ha, while Greece consumes 157 kg NPK/ha, and Jordan has a peak of 682 kg NPK/ha (http://wdi.worldbank.org/table/3.2). Reduction of fertilizer use without losing plant productivity will favor environmental performance, increase nitrogen use efficiency and reduce GHG emissions.

Although action plans and directives (e.g. WFD) to reduce water pollution are being enacted by the EU and by national and regional administrations, many environmental issues are still present. In order to reduce the use of N-P-K mineral fertilizer, the EU-27 policies are also stimulating organic agriculture. EU organic farming areas sum up to 9.6 M hectares representing the 5.4% of the entire EU agricultural area. SME's prospective is that the reduction of N fertilization is expected to reduce farmers production cost, while for policy makers, the need of alternative ways of increasing N use efficiency is imperative to limit pollution issues. A great challenge therefore is to provide farmers with efficient strategies to limit fertilization and water needs.

Important efforts are being addressed by precision agriculture research in the direction of decision support tools for farm management with the objective of optimizing productions while preserving resources [2]. In this contest, models can give support to identify, among alternative choices, those that will not lower the integrity and sustainability of ecosystems. Earth observations supported by model prediction can help stakeholders and end-users involved in earth sciences to support global efforts to understand the physical environment.

The earth science community is generating Big Data that are now being shared on secure networks. The challenge is to extract meaningful information from petabytes of geo-data stored on a variety of data formats and resolutions. This is challenging the community that needs to handle and integrate smoothly such data and to use it to face issues such as food security and production, crop quality and quantity issues, or climate change impacts and mitigation.

2 The Earth Observation System

Earth observation products make use of sensors, advanced mobile devices, pervasive high-speed connections, along with new paradigms emerging from the Internet of Things. Our aim is to contribute with a innovative web-based Earth Observation Product (EOP) for the agricultural and environmental domain that exposes interactive reporting, visualization, and analysis tools. The challenge is to enable the break-through of new technological paradigms and to strengthen the capacity to use and assemble such systems and technologies. Through the set up of a micro-service ori-ented infrastructure, where motivated scientists can co-work, co-design and test applications, we expect to enhance the use of field data and hydrological/climate change models directly on the web to improve the way decision makers observe and manage our planet. E-science can offer high performance computational and data management infrastructure, analysis and visualization tools such as GIS, databases, etc. By working with various groups such as farmers, universities, research centres, and industries we are challenged to bridge the gap between science and application merging in one environment data processing, storing and sharing, modeling, field and remote sensing observation techniques, to which technological development is tightly bound.

For this purpose, a spatial data infrastructure (SDI), connected to a modeling and almost real-time data processing environment is being developed. The SDI is based on a PostGIS relational database management system (RDBMS). The data model adheres to SWAT (Soil & Water Assessment Tool) [4] I/O standards. SWAT is a watershed scale model developed to quantify the impact of land management practices in large, complex watersheds. SWAT geo-data model is organized on 3 different levels: (1) Watershed, (2) Subbasin and (3) Hydrological Response Units (HRU's) (Fig. 1).

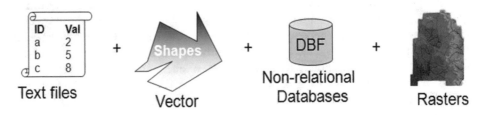

Fig. 1. SWAT I/O is made of geo-data and various ascii files.

Each model implementation (one for each test site) is run for different scenarios, that reflect a change in the input, producing formally an identical database as output. In the back end, each scenario is then processed to be used by the web applications exposed (Fig. 2).

ETL (Extract, Transform, Load) procedures have been developed to process SWAT input and output and produce self consistent relational database instances within the SDI. Both input and output files are stored in the Database.

Each model is run on a High Performance Computing (HPC) environment par-ticularly designed for large distributed data-intensive applications based on SWAT.

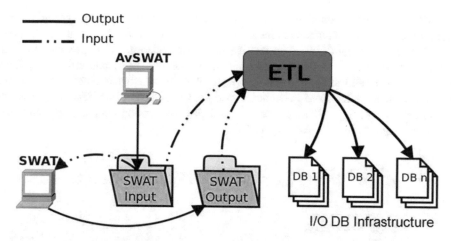

Fig. 2. SWAT Flow Chart. Each Scenario (I/O) is stored in the DB infrastructure.

Datasets, visualization tools, numerical applications and analysis tools are exposed by a web portal that acts as the front-end. The EOP is a scalable analytical frame that allow frequent data input and updates (on climate, soil water content, fertilizer applications, etc.). Data is collected from more than one source (regional climate data providers, field sensors, monitoring networks, EU projects, etc.), analyzed, and fused to create new services displayed on user-friendly interfaces.

Interoperability is of paramount importance in our design. Our interoperability layer is called to transparently share data, computing resources, widgets such as maps, graphs, tables and applications through API and web services.

2.1 The EOP Technological Layer

From the technological point of view, the EOP is based on the Alto framework (www. altoframework.com) [3], a full-stack, modular, visual development framework to design and develop enterprise web information systems, services, and dynamic applications for advanced analysis and reporting. Alto exposes a suite of tools to write server and client side code (HTML5, Velocity, CSS3, and JavaScript). It supplies an abstraction layer between the archives, the query system to the database and the actual reporting mechanism (e.g. geo-processing, widgets engines, etc.). Through Alto the work flow pipelines and the technologies used are transparent to the user that has no knowledge of the complexity of the infrastructure.

The software exposes:

- a user friendly web editor with a powerful GUI.
- a fast and flexible processing system for application development.
- a suite of widget engines to create complex HTML objects such as GIS maps, tables, graphs, search engines, forms, etc.
- a distributed Collaborative Working Environment where large communities can cooperatively create Web applications.

Alto transparently combines, in one space, technologies to access, query and process complex data infrastructures, thus increasing the interoperability between applications.

Alto new paradigm is based on an integrated and collaborative approach where the complexity of the technology is transparent to the end user, and interdisciplinary working groups and skills can be enhanced. Within an experimental programming environment, modules have been developed to run real-time applications based on numerical solvers (e.g. SWAT), run pre- and post-processing codes, query and map results through the web browser (Fig. 3).

EOP Portal HW Infrastructure

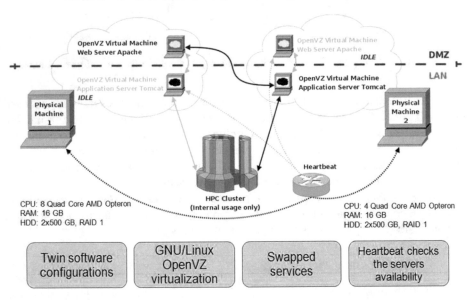

Fig. 3. EOP Portal HW Infrastructure. Linux environment is used to run the SWAT model and the geo-processing phases required by the environmental applications for the reporting production.

A Linux environment is used to run the SWAT model and the geo-processing phases required by the environmental applications for the reporting production. Data flow, data storage procedures, and the application workflows have been developed and recurring operations automated so as to hide the complexity of the infrastructure and limit human made errors.

Our EOP is more than a mere sum of modules: it is a software to consume and expose complex applications, API and web services for data mapping, querying and sharing, processing and distributing, with a high degree of freedom.

The front-end of our EOP exposes results from the application of the physically based SWAT model to assess hydrological, biogeochemical and crop growth cycles. Our platform is expected to enable stakeholders and public end-users to access and use information stored in the SDI on a free, open and easy to understand basis (Fig. 4).

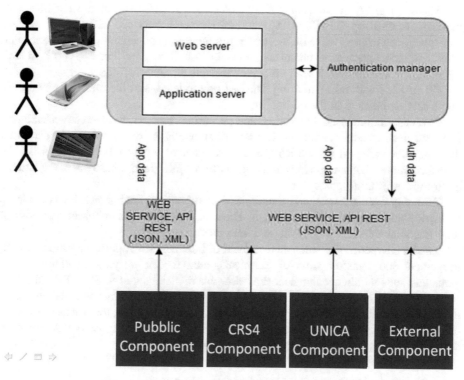

Fig. 4. EOP infrastructure is based on microsrvice.

2.2 The SWAT Model

Consideration of the physical processes connected with water movement, plant growth, and nutrient cycling are decisive to evaluate the build-up of pollutants due to alternative land management practices on downstream water bodies (e.g. rivers, aquifers and coastal lagoons). Soil stores and controls the release of nutrients and their cycle. Bio-geochemical processes within soils are drivers for nutrients to be transformed for plant uptake. Soil controls drainage, flow and storage of water and solutes, which includes nitrogen, phosphorus, pesticides, and other compounds dissolved in the water. With proper soil functioning, plant roots and soil microorganisms interact mutually through diverse physical, chemical, and biological processes, thus supporting soil biodiversity and its habitat. Soil has the ability to maintain its porous structure to allow passage of air and water, withstand erosive forces, and provide a medium for plant roots. Therefore, soil minerals, water, nutrients, microorganisms and crops are the major component of the production equation and pressing questions are asked to drive strategies towards solutions.

The Soil and Water Assessment Tool (SWAT) is a semi distributed watershed-scale model developed by the USDA Agricultural Research Service (ARS) [4, 5]. SWAT was developed to predict the impact of environmental management practices on water, sediment, and agricultural chemical yields in complex watersheds. SWAT models hydrology, sediment movement, crop growth, and nutrient cycling.

Watersheds with no monitoring data can be modeled, and the effect of changes in input data is reflected in the output.

SWAT takes into account weather, soil properties, topography, land cover/land use, and land management conditions. The water balance is the driving force. In particular the fate and transport of nutrients and pesticides in a watershed depends on the transformations that compounds undergo within soils (land phase) and the stream environment (routing phase).

Within SWAT a watershed is subdivided in subbasins. Each subbasisn is made up of Hydrologic Response Units (HRUs). These are portions of a subbasin that possess unique combination of slope, soil, and land use/management.

This subdivision of a watershed reflects differences in evapotranspiration, runoff, movement and transformation of chemicals, etc., for the varying vegetation/crops, slope and soils within a watershed. Processes are first simulated in SWAT for the land phase at the HRU spatial unit. This yields water, sediment, nutrient, and pesticide loadings to the main channel in each subbasin. In a second phase, the water, sediments, etc. are routed through the channel network of the watershed to the outlet. For each subbasin there is one reach, one outlet, and many HRUs.

2.3   Nutrient Cycles and the Crop Growth Module

Phosphorus and Nitrogen within SWAT are modeled in a similar manner as in the Erosion Productivity Impact Calculator (EPIC) model [6].

Nitrate nitrogen quantity in runoff is simulated considering the top soil layer with a maximum contributing thickness of 10 mm. In detail Nitrate−N is calculated as the product of the runoff volume and the nitrate concentration in the first layer. The other soil layers do not contribute to the balance.

Load of Nitrate−N contained in lateral subsurface flow and percolation are estimated as products of the water volume and the average concentration of nitrate in each layer. SWAT implements to compute Organic N transport with sediment using the loading function developed by McElroy et al. in 1976 [7] and modified by Williams and Hann in 1978 [8].

In the top soil layer, the loading function estimates the daily organic N load within runoff based on the concentration of organic N the sediment yield, and the enrichment ratio. The enrichment ratio is the coefficient of the mass of organic nitrogen in the sediment to that in the soil. Moreover, plant uptake of nitrogen is calculated using a supply−and−demand approach (Fig. 5).

Given that phosphorus is scarcely soluble, phosphorus loads transported in surface runoff is computed using the schematization as described by Leonard and Wauchope (1980) [9]. In particular, the amount of soluble phosphorus taken in runoff is calculated using the labile P concentration in the first 10 mm of the soil, the runoff volume, and a partitioning coefficient. The transport of P (particulate P) within sediment is estimated

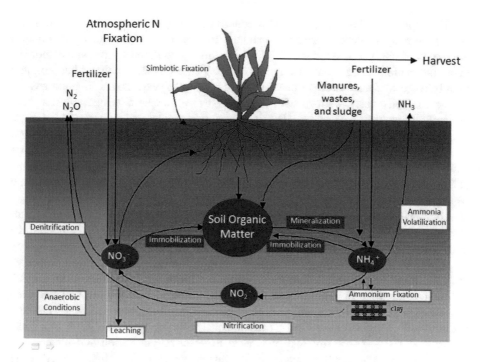

Fig. 5. Nutrient Cycle in SWAT: N

using a loading function, as described above in organic N transport. Phosphorus taken by crops to grow is estimated with the supply–and–demand approach.

SWAT has integrated a simplified version of the QUAL2E model [10] to take into account in–stream nutrient transformations. QUAL2E is steady state Enhanced Stream Water Quality Model and within SWAT is used to simulate conventional pollutant transformation in branching streams and well mixed lakes. The sub–components of QUAL2E integrated into SWAT include biochemical processes such as algae, chlorophyll–a, dissolved oxygen, carbonaceous oxygen demand, organic nitrogen, ammonium nitrogen, nitrite nitrogen, nitrate nitrogen, organic phosphorus, and soluble phosphorus.

Crop growth is modeled by SWAT using the Potential Heat Unit (PHU) approach. This is an implementation of the EPIC model within SWAT to simulate and quantify biophysical process related to plant growth. PHU are the number of heat units required to bring a plant to maturity.

All model outputs are given at the subbasin/HRU level at a daily temporal scale.

3 The Sardinian Case Study

In this study, the SWAT model is used to assess agricultural draught and the impact of agricultural practices on water quality of the 520 km^2 S. Sperate basin (South West of Sardinia, Italy).

The climate of the area is typical Mediterranean with hot long dry summers and short mild rainy winters. Average temperature ranges from 8 °C (January and February) to 25 °C (July and August). Rainfalls are largely confined to the winter months, being the rainfall regime characterized by a peak rainfall in December and a minimum in July/August. Yearly average precipitation is 591 mm/year. The S. Sperate river is characteristically fast flowing, with higher water volumes in winter, almost non-existent during the dry season. The monthly water volume is characterized by a minimum peak in August (0,16 mc/s) and a maximum in February (4 mc/s).

Land is primarily used to satisfy agricultural needs with large areas destined to crop cultivation (Cereal is predominant – 9091 ha). On the south we find vineyards (1709 ha), olive groves (2383 ha) and orchards 1709 ha) mainly. At East, woods and pastures are mostly found.

High levels of P, NH4, NO3 and COD, etc. are found in the two monitoring gages located on the San Sperate river within the basin. Such high values are assumed to be mainly due to the agro-zootechnical compartment.

The hydrological behavior of soil is related to a number of physical soil properties [11]. To obtain information about these soil properties of the S. Sperate basin, a 1:250 000 soil vector map [11, 12] has been used, where each cartographic unit has been associated with one or two delineations corresponding to subgroups of USDA soil taxonomy. Vegetation cover significantly affects the water cycle. The CORINE Land Cover [13] 1:100.000 vector map was used to describe vegetation and land use. It consists of a spatial database made of 44 classes, grouped into three nomenclature levels. that covers the entire spectrum of Europe.

Model input data along with the watershed and HRU spatial discretization criteria [14, 15] were carefully checked to ensure global consistency.

Simulation cover over the period 1922–2008. Temperature and precipitation daily data for the period January 1995–February 2008 were used in the simulation, recorded by monitoring network of the Regional Environmental Agencies for Sardinia (ARPAS). The climatic stations used for this application are located within the watershed or localized in the surroundings: Decimomannu, Dolianova, Guasila, Siurgus-Donigala and Villasalto. Figure 6 shows the location of the San Sperate watershed and the location of 3 monitoring stations (Fig. 7).

The SWAT model has been calibrated using the SWAT CUP software. Monthly stream flow values were used in the calibration process, scoring 0.73 of the Nash Sutcliffe index with a correlation factor of about 0.83 showing a good accuracy in reproducing the hydrologic regime.

We show (Figs. 8 and 9) the impact of point and diffuse pollution on water quality. The exploited land is predominantly destined to cereal, vineyards, olive groves and orchards. In the area, for example cereals are being fertilized with: (i) 120 kg/ha 18-46-00 fertilizer applied on mid December and (ii) 60 kg/ha 46-00-00 fertilizer applied on mid February. The eastern portion of the watershed is predominantly forest and pastures (about 75000 ovine and 4500 bovines). Four treatments plants are also present and route their water directly on the river.

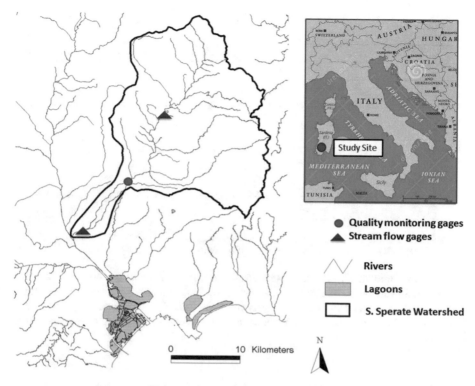

Fig. 6. The 520 km^2 S. Sperate basin (South western part of Sardinia, Italy)

Both diffuse and point pollution sources have been accounted in the model.

A methodology to evaluate agricultural drought conditions has been also set up. Drought is a temporary condition of relative scarcity of water resource compared to values that can be considered normal for a period of time and on a region. We may distinguish between meteorological, agricultural, hydrological and operational drought. While the meteorological drought is identified on the basis of a deficit of precipitation, the agricultural drought depends on the soil moisture deficit, which is dependent on many factors such as the precipitation regime and weather, the soil characteristics and the evapotranspiration rate. The persistence of agricultural drought condition produces negative effects both on natural vegetation and agriculture. Drought periods have an important impact on water supply system causing water shortage, negatively affecting the economic and social system.

In this study, we implemented a variation of the approach proposed by Narasimhan [16] to calculate the Soil Moistures Deficit (SMD) agricultural drought index. SMD at the sub-basin spatial scale is computed on a monthly basis as proposed in the formula (1). For the given month the index expresses the ratio between the anomaly of the monthly value compared to the average multi-annual data, and the difference between the maximum and minimum values for the entire time series available.

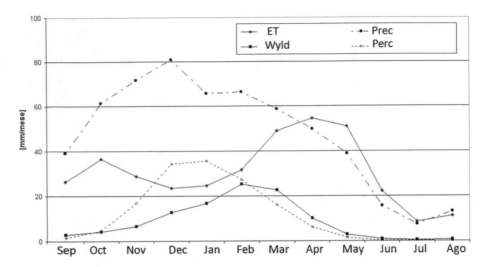

Fig. 7. Prec - Rainfall, ET - Actual Evapotranspiration, Perc – Percolation, Wyld – Water Yield for the San Sperate basin. Mediterranean summers are long and dry. The warm season starts in May and ends in September. Winters are very mild and wet. Very few places experience snow. The seasonal changes are controlled by the Mediterranean and its change in temperature during the year.

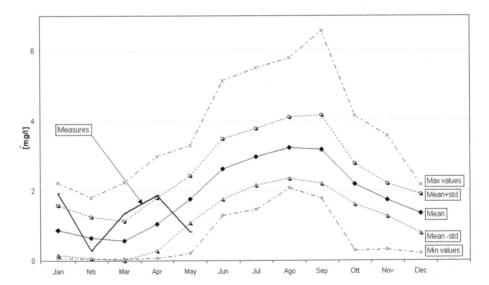

Fig. 8. Simulated NO3 concentrations and the measured concentrations for the 20801 gage (outlet of the watershed).

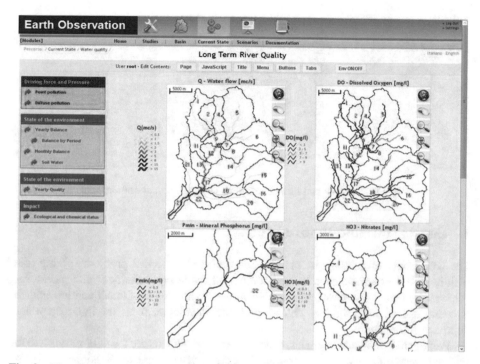

Fig. 9. Visualization of SWAT output in the web interface. SWAT has been employed to quantify the impact of land management practices on rivers. Spatial analysis of simulated long term average Q (stream flow), Dissolved Oxygen, Nitrates, Mineral Phosphorous.

The SMD index is represented by the following formula:

$$SMD_i = \frac{SW_i - SW_i^{mean}}{SW_i^{max} - SW_i^{min}} \tag{1}$$

Where:

SMDi represents the deficit of soil water content of the i months,
SW_i the monthly average soil water content of month i,
SW_i^{mean} the long-term average of the soil water content of month i,
SW_i^{min} and SW_i^{max} respectively the minimum and the maximum soil water content of month i for the entire simulation.

The index can have positive or negative values, signifying for a given soil and for a given month a surplus and a deficit of water content respectively. The anomaly magnitude of the SMD drought index, mediated on each month and on a subbasin spatial scale represents drought. In Fig. 10 for March 2007, we provide the spatial distribution of the monthly SMD index. Yellow (SMD < –0.1) and orange (SMD < –0,3) colors represent area under water stress while green colors show high water content values.

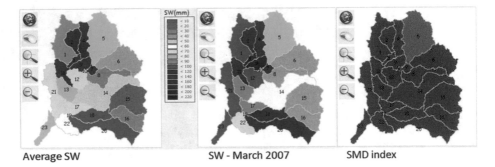

Average SW   SW - March 2007   SMD index

Fig. 10. Yellow and orange refers to extreme agricultural drought.

4 Conclusions and Future Work

The complexity of natural systems requires the use of efficient problem solving tools mainly based on remote sensing, environmental modeling techniques and monitoring networks. In such a contest, reliable decision activities needs the acquisition and the efficient use of large quality dataset and the development of an interdisciplinary approach to the study.

Public environmental agencies, stakeholders and citizens are steadily moving to the Internet, searching ever more for reliable data, analysis and applications. Although monitoring, studying and understanding environmental dynamics are highly costly, time consuming activities and require an integrated approach, cooperation is still limited. This results in inefficiency in the decision process.

Environmental sciences are evolving from a simple, local-scale approach toward complex multilayered, spatially explicit regional ones. Large amount of geo-data now available (see Copernicus project [17]), increased needs of end users, new challenges set by emerging technology are opening new grounds of discussions. The advances in computer science are extending the possibilities in earth sciences, and are deeply changing the ways in which observation and management systems can operate, providing sophisticated analysis facilities. The new envisioned paradigm is based on collaborative/integrated web based approach in which the complexity of the technology is transparent to the end user, and interdisciplinary working groups and skills can be enhanced. In the Earth Science context, scientists can now transparently access, through scientific portals, to distributed data repositories (across several domains and institutions), metadata sources, and perform search and discovery activities.

An increased attention on the interoperability between data and service providers for the sharing of knowledge is expected also to improve data interpretation abilities in the future.

The use and integration of advanced ICT technologies and open standards on the web still involves major investments, and alternative technological solutions can be adopted. E-science can offer high performance computational and data management tiers providing a significant contribution in the description and management of water and soil resources.

With this work we applied a modeling infrastructure to provide farmers and public and private end users with an integrated and participatory support framework and specific analytical tools to support a more sustainable crop production. The use of modeling tools can help reducing the use of N/P fertilizer and water and evaluate conditions to sustain growth of seasonal crops and plants under different climate and soil conditions. The use of such analytical framework can improve management practices, the monitoring and characterization tools to assess soil biodiversity and rhizosphere mineral evolution to reduce water and nutrient (N/P) need for growing crops.

The proposed procedure for estimating the SMD drought index is based on the hydrological balance calculated by the SWAT model on the HRU spatial scale and on the daily time step. Such analysis is done at a correct scale considering the hydrological phenomena involved. Subbasin monthly estimates are, in fact, derived from the daily hydrological budgets.

Future farming can only improve decision-making and farm management if farmers have access to the necessary scale, detailed information to make rationale choices. The creation of model based decision at the field or even plant-level can support farmers to improve their crop production and plan long-term investments. We aim at producing a more comprehensive and up-to-date almost real time picture benefitting from GPS technologies (in the future also drones will be considered), to better decide as to where and when to apply irrigation, how much fertilizers to give, when to harvest and so forth.

We expect to improve model usability and the ways in which land management systems can operate to aid in making management decisions and watershed-scale modeling.

Further work will deal with the identification of the factors that drive agriculture production and its change, including environmental drivers (notably but not exclusively climate and soils), policy drivers (notably but not exclusively regulation and incentives), and market drivers (notably but not exclusively prices and costs).

Various actors have expressed the need to bridge the gap between natural, physical, policy, economic and social science and data that underpin the understanding of the interplay of these factors in changing agricultural contexts. Identifying the effects of site specific constraints upon drivers of land management's, and their impacts is complex but can be achieved. The use of model based decision support system can help develop management strategies that include biophysical impacts on the environment, and distribution of impacts across stakeholders.

Acknowledgment. This work has been partially funded by the Top-down TESTARE project, the EranetMED SUPREME project and the Regione Autonoma della Sardegna (R.A.S.).

References

1. Tilman, D., et al.: Proc. Natl. Acad. Sci. USA **108**, 20260–20264 (2011)
2. http://docs.opengeospatial.org/wp/16-131r2/16-131r2.html#_ftn7. Accessed 21 Sept 2018
3. www.altoframework.com. Accessed 21 June 2018

4. Neitsch, S.l., Arnold, J.G., Kiniry, J.R., Williams, J.R.: Soil and Water Assessment Tool. Theoretical documentation. Version 2000. Blackland research center - Texas agricultural experiment station. Grassland, soil and water research laboratory - USDA agricultural research service (2001)
5. USDA & NRCS Soil Survey Division: State Soil Geographic (STATSG0) Data Base - Data use information. Miscellaneous Publication Number 1492 (1994)
6. Williams, J.R.: Chapter 25. The EPIC model. In: Computer Models of Watershed Hydrology, pp. 909–1000. Water Resources Publications, Highlands Ranch (1995)
7. McElroy, A.D., Chiu, S.Y., Nebgen, J.W., et al.: Loading functions for assessment of water pollution from nonpoint sources. EPA 600/2−76−151. USEPA, Athens (1976)
8. Leonard, R.A., Wauchope, R.D.: Chapter 5: the pesticide submodel. In: Knisel, W.G. (ed.) CREAMS: A Field−Scale Model for Chemicals, Runoff, and Erosion from Agricultural Management Systems, USDA Conservation Research Report No. 26, pp. 88−112. USDA, Washington, D.C. (1980)
9. Williams, J.R., Hann, R.W.: Optimal operation of large agricultural watershed with water quality constraints. Technical report No. 96. College Station, Texas A&M University, Texas Water Resources Institute, Texas (1978)
10. Brown, L.C., Barnwell Jr., T.O.: The enhanced water quality models QUAL2E and QUAL2E−UNCAS documentation and user manual. EPA document EPA/600/3−87/007. USEPA, Athens (1987)
11. Arangino, F., Aru, A., Baldaccini, P., Vacca, S.: I suoli delle aree irrigabili della Sardegna (Piano Generale delle Acque, Regione Autonoma della Sardegna), Cagliari (1986)
12. Aru, A., Baldaccini, P., Vacca, A.: Nota illustrativa alla carta dei suoli della Sardegna, con carta illustrata 1:250.000. Regione Autonoma della Sardegna (1991)
13. https://land.copernicus.eu/pan-european/corine-land-cover. Accessed 21 June 2018
14. Cadeddu, A., Lecca, G.: Sistema Informativo dei Suoli della Sardegna da interfacciare al modello idrologico SWAT. Technical report 09/03. CRS4, Center for Advanced Studies, Research and Development in Sardinia. Cagliari, Italy (2003)
15. Cau, P., Cadeddu, A., Gallo, C., Lecca, G., Marrocu, M.: Estimating the water balance of the Sardinian island using the SWAT model. Published in the Journal "L'Acqua" N. 05, September–October 2005, pp. 29–38 (2005)
16. Narasimhan, B., Srinivasan, R.: Development of a soil moisture index for agricultural drought monitoring using a hydrologic model (SWAT), GIS and remote sensing. In: Texas Water Monitoring Congress, 9–11 September 2002, Austin, TX (2002)
17. http://www.copernicus.eu/. Accessed 21 Aug 2018

Electronic Crop (e-Crop): An Intelligent IoT Solution for Optimum Crop Production

V. S. Santhosh Mithra$^{(\boxtimes)}$ (iD)

ICAR-Central Tuber Crops Research Institute, Sreekariyam,
Thiruvananthapuram, India
vssmithra@gmail.com

Abstract. Electronic Crop (e-Crop) is an electronic crop simulator. The device computes how much food is produced by the plant using the given sunlight, water and other factors. It informs the farmer about the status of the crop, its input requirements to realize targeted yield as well as about the forecasts regarding the crop and scheduling of irrigation, nutrient applications, agronomic and plant protection operations. Forecasting of yield of the crop can be done more accurately at local, regional and national level. Device gives information to the farmer in the form of SMS. This device can be used for giving real-time agro advisory on any crop to reduce yield gap and to achieve targeted yield. One device is sufficient for different crops grown in a continuous geographical area with uniform weather conditions. It is a single point solution for many problems in agriculture. The main advantage of this IoT device is that it retrieves real time information about the various soil and weather parameters for analysis and thereby giving the respective farmer with precise advisory.

Keywords: IoT · Automation · Simulation · Forecasting · Agrobotics
Modelling

1 Introduction

Present day food production is not sufficient to achieve zero hunger especially in the third world countries where food demand is increasing continuously for the ever increasing population. Since the cultivable area is shrinking in this part of the world, achieving higher productivity from the available genetic stock is the only solution. With the help of computer simulation models, it is identified that the difference between potential and achieved yield is generally wide and narrowing this gap is the key strategy to meet the increasing food requirements of the future [1]. Attending the individual crops and applying required inputs in the exactly sufficient quantity leads to increase in production and minimize the loss of these chemicals to the environment [2]. This helps to conserve the natural environment and preserves its potential for producing more from it. Revolutionary inventions are happening in many faculties of science like ICT, space science etc. Precision farming by integrating these technological advances

---

V. S. S. Mithra—Principal Scientist (Computer Applications).

J. C. Corrales et al. (Eds.): AACC 2018, AISC 893, pp. 177–189, 2019.
https://doi.org/10.1007/978-3-030-04447-3_12

of information age is the best strategy to reduce this gap [3]. Advisories on plot specific management, after assessing remote sensing pictures of the crop and field are already in use [4]. Forecasts made in this way are prone to error [5, 6] mainly because the key factors like crop physiology are not considered. By pooling such data regional and national level crop forecasting are being done in many countries [7]. To support the precision farming strategies, ICAR-CTCRI has developed an electronic crop simulator, which communicates to the farmers directly from the field on behalf of the crop and inform him about the status, requirements and what will happen to the crop in future etc. This device is named Electronic Crop or e-Crop in short.

A lot of web based agro advisory services are already prevailing in many parts of the world and for many crops [8, 9]. Many of the agro advisories are to manage their crops for better yield [10] or for some special solutions like pest and disease control [11, 12]. Most of the advisories are not accurate. But accurate, precise and timely advisory is always in demand. This paper describes Electronic Crop (e-Crop) which simulates the crop growth using the weather and soil data collected real-time from the field directly.

This device was developed on the sound footing laid by the progress made in the field of crop simulation modeling. Crop model is a simple representation of the crop [13]. It is used for simulating the physiological processes happening in the crop in response to the environment. Models which are based on sound physiological data are capable of quantification of temporal and spatial variability and can be very useful for precision farming [14].

1.1   SPOTCOMS Model

First e-Crop was developed to simulate the growth of sweet potato. SPOTCOMS model [15] was used to represent physiology of the crop. Growing degree days (GDD) was used for computing phenological development. Weather data parameters like maximum and minimum temperature (°C), solar radiation (MJ m^{-2}day^{-1})/sunshine hours (h), precipitation (mm) and plant parameter values are the major driving variables of the SPOTCOMS model. Nutrient parameters like applied and available doses of N, P$_2$O$_5$ and K$_2$O are also the necessary inputs of the model. The model calculates the plant attributes like vine length, number of leaves, leaf area index, number of tubers, number of branches and dry matter in roots, tubers, leaves and vine. Dry matter accumulated in tubers at different levels of moisture, N and K$_2$O are also calculated by the model. Software of the SPOTCOMS model was modified so that it can be used in the device where the input files are read automatically from the directory.

2   e-Crop

This is a weather proof electronic device, developed for working in any terrain under various climatic conditions (Fig. 1). The device consists of a main control unit that accepts various real-time parameters of soil and weather conditions and different sensors which are used to read different data of soil and weather. The micro controller inside the control unit is responsible for coordinating the clock, initiating the internet connection and also for the data collection from different sensors.

Fig. 1. View of e-Crop installed in the field at ICAR-CTCRI, India

First e-Crop unit was developed to simulate the sweet potato growth in response to the driving variables without any human involvement. This is a weather proof electronic device which works directly in the field. Design of the device is given in Fig. 2. Sensors are used to collect data on weather and soil parameters. The data collected by the device are (a) Temperature (b) Humidity (c) Solar radiation (d) Soil moisture (e) Wind velocity (f) Wind direction (g) Rainfall and (h) Geo coordinates.

The data collected by the sensors are sent to Raspberry Pi (RPI) for processing. RPI sends this data to the cloud at 15 min interval. Sensors are positioned on the exterior of the box and are interfaced to the motherboard using RJ-11 jacks.

2.1   Sensors of the Device

Sensors used in the device are: (1) Temperature and Humidity sensor (2) Solar radiation sensor (3) Soil moisture sensor (4) Wind speed and wind direction sensor and (5) Rain gauge sensor.

Temperature and Humidity Sensor: Sensor used for the measurement of temperature is TRH-304. This sensor is a multiplexed sensor with the capability to measure both temperature and humidity. TRH-304 series uses a thin film polymer capacitor to sense relative humidity and a thin film RTD to accurately sense temperature. It is a current loop sensor requiring supply of 12–30 VDC with output range of 4–20 mA, of which 4 mA indicates 0 °C and 20 mA indicates 100 °C. From this, we can calculate current per °C.

Solar Radiation Sensor: Sensor used for measuring radiation is pyranometer. We use Apogee SP-215 s: Amplified 0–5 V Pyranometer. The SP-215 is an amplified sensor with a 0–5 V output and exhibits excellent cosine response. All Apogee pyranometers

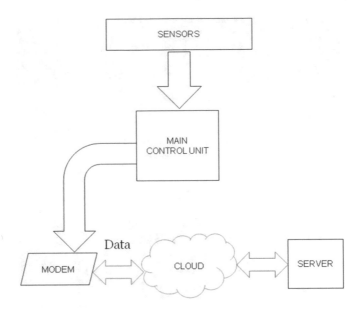

Fig. 2. Schematic diagram of the device

incorporate a silicon-cell photodiode that measures total shortwave radiation with a sensor housing design that features a fully potted, domed-shaped head making the sensor fully weatherproof and self-cleaning. Total shortwave radiation is an important component in determining evapo-transipration rates, energy balance, net radiation as well as monitoring solar power panels.

The output from SP-215: Amplified 0–5 V Pyranometer is a DC voltage ranging from 0–5 VDC. SP-215: Amplified 0–5 V Pyranometer has a calibration factor of 0.25. The output solar radiation is converted to W/m^2.

The e-Crop motherboard system works on 3.3 V. As our sensor is of 5 V range. we need to convert 5 V level to 3.3 V. The same can be achieved by using a voltage divider bias.

Soil Moisture Sensor: The sensor used for the measurement of soil moisture is AT210. The AT210 is a current loop sensor with output range of 4–20 mA, of which 4 mA indicates 0% and 20 mA indicates 100%.

Wind Speed and Wind Direction Sensor: The cup-type anemometer measures wind speed by closing a contact as a magnet moves past a switch. A wind speed of 1.492 MPH (2.4 km/h) causes the switch to close once per second. The anemometer switch is connected to the inner two conductors of the RJ-11 cable shared by the anemometer and wind vane.

Rain Gauge Sensor: The rain gauge is a self-emptying tipping bucket type. Each 0.011″ (0.2794 mm) of rain causes one momentary contact closure that can be recorded with a digital counter or microcontroller interrupt input. The gauge's switch is connected to the two center conductors of the attached RJ-11 terminated cable. A rain fall of 0.2794 mm will cause the sensor to produce a tip/pulse. The tip is counted in the mother board.

2.2 Working of the Device

Schematic overview of working of e-Crop is shown in Fig. 3. The device works for any crop, pest, nutrient etc. provided its simulation model is loaded for analyzing the data collected by the sensors. Step by step working of the device is:

- Install the device safely in the field in a solid platform. The solar panel also should be placed near the device and the wires should be well protected against biotic and abiotic damages.
- Data accession is done on demand from the Raspberry Pi. A command from the Pi is replied by the mother board with the current measured parameters at that instant. The time is controlled by Raspberry Pi. Time for updating of parameters are set to 15 min. The processor in the mother board runs on 3.3 VDC. The voltage is selected so to ease the communication between the Raspberry Pi and mother board. The Raspberry Pi runs on 3.3 VDC and a voltage above 3.5 VDC will cause serious damage for Raspberry Pi.
- The sensors are activated every 15 min to read the data and send it through cables which are interfaced to the main control unit using RJ-11 jacks. A modem is integrated to initiate the internet connection. The data that is being collected from

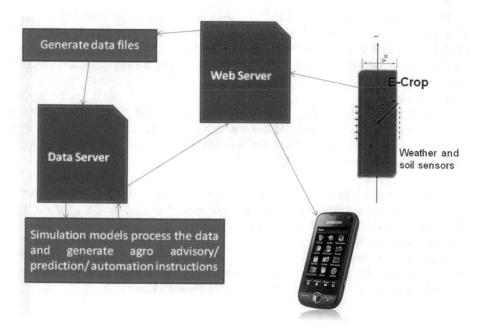

Fig. 3. Schematic overview of the working of e-Crop

the sensors are then uploaded, using the internet connection available, to the server in a remote location.

- At the end of the day, these weather data are converted into the format specific to the models, which use these data as driving variables. The data files will be downloaded to the server where the models reside using a scheduler software.
- The concerned simulation model(s) gets executed using these data files and generates agro advisory/prediction.
- These advisories is sent to the cloud from where these are sent to the mobile of the concerned farmer in the form of SMS.
- These predictions are sent to the cloud from where these are sent to the server where these are compiled.

A management interface is used to control and communicate with the advanced settings of the main control unit. This web interface resides in the cloud and communicates with the control unit. The main functions of this interface is to add users and their privileges, and inputting various values of crops, fertilizer, soil type, locations, adding new device etc., according to the privileges assigned.

2.3   Web Interface to Configure and Manage the Device

A web interface was developed to manage the device. The interface was developed using PHP/MySQL. Various facilities in the interface are:

User Types:

Farmer

Normal primary user type. Limited rights only. Access to Fertilizer Adding Modules and View Advisory Text (Old and Current). This user type is created by **Device Owner** or **Admin** user from "ECROP USER" Link. The system will generate an access code and send to farmers Email and inform via SMS message to farmers mobile number.

Device Owner

Secondary type of user type. Has limited access to system but has the right to create **Farmer** user, View Sensor data, Add soil type and add Locations under the purview of the given e-Ccrop Device ID. **Device owner** can register them self via "New User Registration" with a new Device ID.

Admin

Admin user is the super user of this system. Has access to all modules and has the right to view all **Device Owner** users and **Farmer** users from Applications. **Admin** user can access the ADMIN MODULES. **Admin** user name is hardcoded to the application with the username "admin".

The first user should be admin when we setup a new installation. The password reset link will be sent to Admin's registered mail. By clicking on this link Admin can reset the password. (This is applicable to **Device Owner** also)

User Login Sections:

Farmer Login Module

This module is for **Farmer** to add fertilizer as well as to view the advisory history. The access code necessary to open these modules has been generated by the **Device owner** or **Admin** while creating the **Farmer** user.

Admin Login Module

Admin Login module is http://ctcri.org/ecrop/login.php. **Admin** user name is hardcoded by login name "admin" and **Admin** user has complete access to all modules.

Admin Modules:

Device

DEVICE LINK is for creating new devices. This modules used to create new device ID, Once we manufacture a new device we have to assign a new device ID to the hardware and a printed ID to the buyer.

Location

This link is used to create new location under a device. We can add multiple locations to a unique devices.

e-Crop Users

Used to create **Farmer** users for e-Crop. **Admin** and **Device Owner** can create a **Farmer** user, by providing mobile number and email along with other details. The system will generate the access code and send to email as well as to mobile number as SMS.

Admin user can view the access code and mobile number of each **Farmer** below the page. A **Farmer** can access the system at "Farmer Login" link from main Page.

Crops:

Used to create new crops from **Admin**. System will ask for crop name, model name and duration. Here the model name is the physical directory name of crop model engine, crop duration is the duration of crop maturity in days. From this page we can create the crop variety too. To create the crop variety please use the "Add Variety" link alongside the existing crops.

"Add Variety" link is used to add new varieties of the crops, System will ask for the name of the variety and its duration.

Soil Type:

Used to add soil type by the **Admin**. This soil type have to be selected by **Device owner** user, while creating a new **Farmer** user.

Fertilizer:

Using this module new fertilizers can be added. To add fertilizers to the crop, **Farmer** can select those fertilizers available in the system, which have been updated using this module. Apart from quantity **Admin** can add the unit also. Default unit is **(kg/ha)**.

Sensor Data:

This module helps to view the raw sensor data collected by the e-Crop device. At present this can be viewed by all users.

Chart:

Chart is the graphical representation of collected data. This is restricted to **Admin** or to **Device owner** users.

#SMS LOG# or Advisory Text History:
Helps to display the advisory text of current date as well as the earlier date of any specific user. Apart from this, the system will store the all SMS related logs for security reasons.

Logout:
With this link user can logout from the session, and clear all user sessions. After logout, the system will be directed to the login page.

3   Discussion

Progress made in weather and climate forecasting ability has opened up the possibility of strategic adaptation of agricultural management in anticipation of weather and climate outcomes [16]. This device can be used for giving real-time agro advisory of any crop to reduce yield gap and to achieve targeted yield. Weather parameters of the day, the potential yield that can be achieved by the crop after its stipulated duration as per its present crop condition and anticipated weather scenarios, N, P, K and moisture required to be applied to achieve this targeted yield etc. are part of the advisory received in the mobile phone as SMS and the farmer can follow these strategies to increase the yield to the desired level. Such appropriate diagnostic tools that help in the application of fertilizers at the time of demand and in smaller and frequent doses can help to reduce the losses while maintaining or increasing the yield from the crops [17]. This is an excellent device for precision farming which collects the data real-time from the field and then generate advisory and inform the farmer about the present and future status of the crops as well as the strategies to manage the crop to get better results. The data collected by the devices installed in different fields give a very clear realistic overall status of the crop at present and in future. This information will be useful to the policy makers and planners as well as for averting the market risk which usually emanates from unexpected boom in production/supply, fall in prices etc. If information about the production is known well in advance, sufficient precautions can be taken to avoid such risks.

3.1   Field Testing of e-Crop

The device was installed in the crop museum of ICAR-Central Tuber Crops Research Institute (ICAR-CTCRI), Thiruvananthapuram, India for testing the accuracy of its predictions. Sweet potato simulation model SPOTCOMS made predictions on potential yield and actual yield on each day. The planting was done on first week of March, 2016 i.e. on March 8, 2016. The average weather conditions of the crop growth season are given in Table 1.

The predictions were made during the entire growth season (Table 2) and then it was tested against the tuber yield at final harvest and it was observed that the predicted and observed values showed good agreement. Advisories of e-Crop was not followed to reduce the yield gap and hence the potential yield predicted came down each day and at harvest both potential and actual yields became almost the same.

Table 1. Average weather parameters during the crop season March-June, 2016 at ICAR-CTCRI, Thiruvananthapuram

Mean temperature (°C)	Solar radiation (MJ/m²/day)	Wind velocity (Km/hr)	Precipitation (mm/day)
31.09	24.14	16.07	0.10

Table 2. Table showing yield predictions made by e-Crop

Date of prediction	Potential yield (t ha^{-1})	Actual yield (t ha^{-1})
15/03/2016	15.00	12.97
31/03/2016	13.37	12.98
15/04/2016	13.18	12.97
25/06/2016	13.06	12.97

Harvest was done on 25/06/2016 and the predicted and observed values of tuber yield on that day were 12.97 and 12.15 t ha^{-1}. These values show that e-Crop's prediction was 94.25% accurate.

3.2   Problems Solved by e-Crop

Productivity levels in third world countries like Indian agriculture is very low as compared to that of other countries. Difference between achievable and achieved yield i.e. yield gap is generally wide under Indian conditions. Breeding for genetically improved varieties to achieve higher productivity as well as application of higher dose of water, nutrients etc. to reduce yield gap are the common strategies to achieve higher productivity. Long time delay happens in getting results from breeding work to produce better genetic stock. Over application of water and nutrients deteriorates environment and the cost of cultivation increases. The e-Crop helps to achieve higher productivity by reducing yield gap as:

- This product daily calculates plot by plot yield gap and quantifies, N, P, K and water requirement to reduce it
- This information is sent to owner of the plots daily as SMS
- Through the daily/frequent application of nutrients and water, its total requirement for the entire season is less (about 25–50% reduction) whereas yield increases at least by 100%
- Reduced application of the chemicals and water, save resources and minimizes damage to environmental
- Farmers profit multiplies by the increase of yield as well as by the lowering of cost of cultivation

Crop yield as well as pest and disease forecasting and agro advisories in many countries are done more or less at macro level. These forecasting and advisories are done using satellite images, crop cutting experiments, weather data etc. Most of the cases the forecasts are less accurate though large areas can be covered by satellite images. With satellite images micro level data/prediction are not accurate. Crop cutting experiments which are being practiced in many countries like India for crop forecasting are elaborate, tedious, costly and less accurate. The e-Crop solves these problems as:

- The e-Crop predicts crop's yield, plot by plot using the weather and soil data of individual fields and hence predictions are more accurate
- Thoroughly and extensively validated crop simulation models which are developed based on crop physiology are used for predictions and hence it is more reliable
- Advisories on yield prediction and management strategies to reduce yield gap are sent to individual farmer's mobile
- The forecasts sent by the e-Crop devices from different fields can be pooled automatically to obtain yield prediction at national/state/regional level for different times in future with very high level of accuracy. This is an alternate and better solution for present day yield forecasting which is being done using crop cutting experiments at 8500 locations across the nation
- Predictions by pest and disease models installed in e-Crop are more accurate because the predictions are based on the microclimatic data and this helps in better pest and disease control

3.3   Advantages Over the Known Alternatives

Major uses of e-Crop are in the field of **Agro advisory services** and **crop yield forecasting**. This device has the potential to revamp the way in which both these services are functioning today

1. Agro advisory services (AAS): In India this service is provided by National Informatics Centre in collaboration with Department of Information Technology, Government of India. The present AAS is an elaborate multilevel set up involving huge number of human labour, expense, time, resources etc. Data on weather parameters are collected at the Indian Meteorological Department (IMD) weather stations and are compiled at district headquarters or state head quarters. Universities, Krishi Vigyan Kendras (KVK) or Research Institutes are the stakeholders of the data. This data is processed at Agro Metrological Field Unit (AMFU) and agro advisories are prepared. The advisory is sent every Tuesday and Friday to Deputy Director of Agriculture (DDA), KVKs, Krishak Bharati Cooperative Limited (KRIBHCO), Universities, Non Governmental Organizations (NGO's), Agricultural Technology Management Agency (ATMA) etc. Deputy Director, Agriculture (DDA) acts on the advisory through its agencies. KVKs also acts on advisory by displaying in display board and through Mass Media.

    This cumbersome process can be simplified a lot and the quality of the advisory can be improved by using this device. One e-Crop device each can be installed for the continuous area experiencing same weather. From this area, the device collects weather data of all the important parameters like temperature, humidity, solar radiation, wind velocity, wind direction, day length and geo-coordinates like latitude, longitude and altitude. Each device has a unique ID number and the data collected can be sent to a central server along with its geo-coordinates, time etc. There is no need to compile the data manually at any point. At the central station the data from all the devices across the country gets accumulated and this gives a more realistic picture of the weather condition of the country as a whole. Along with the weather data the agro advisory generated by each device for the crops within the range also gets accumulated. There is no need to take any measure to send this advisory to the clients since the device itself does it automatically. By pooling this advisory the crop condition across the country can be known easily and with more accuracy.

2. Crop yield forecasting: In India forecasting of crop yield is done under FASAL (Forecasting Agricultural output using Space, Agro-meteorology and Land based observations) programme of Ministry of Agriculture and Farmers Welfare. Crop forecasts are regularly generated at District/State/National level for major crops using the procedures developed by Space Applications Centre, ISRO. These predictions are done based on high resolution satellite images and ground truth verifications are done at more than 8500 centres. Multiple approaches are followed for different crops.

When we set up e-Crop network throughout the country directly in farmers field, yield forecasting of all the crops can be done more accurately at regional, state or national levels. By pooling the predictions made by each device in a central server, yield forecast of the crops can be made at the desired level. Accuracy and resolution are more when this device is used. Forecasting done with the help of satellite images have limitations which happens when the image capturing is hindered by many obstacles or due to poor resolution of images etc. Such limitations are overcome when this device is used for the same purpose. Since crop simulation models are used and since the weather and soil data are collected directly from the field, the forecasting by this device is more accurate and reliable. Such accurate advance information is helpful to the farmers in many ways especially to formulate strategies to tackle market risks due to glut in the market, low price etc.

3.4   Applications of the Device

- Forecasting of yield of the crop can be done more accurately at local, regional and national level.
- The forecasts sent by the e-Crop devices installed in different fields to the centralized database can be pooled and a national/state/regional level crop yield/status can be obtained by just compiling that information at the desired level for different times in future.
- Device gives information to the farmer in the form of SMS about what is happening to the crop even if he is far away from field.

3.5   How e-Crop Solutions Stands Out from Its Similar Counterparts

- A single point solution to nutrient management, irrigation and plant protection and more solutions like price forecasting, information exchange etc. can be integrated into it
- Very simple and less costly tool for digital farming compared to other methods, where establishment of automatic weather stations, sensor networks etc. are essential
- Easily upgradable since the enhancements/up gradations are mostly done in the software and hence it can be easily incorporated in the system through cloud. Hardware enhancements are not quite frequent.

References

1. Lv, S., Yang, X., Lin, X., Liu, Z., Zhao, J., Li, K., Mu, C., Chen, X., Chen, F., Mi, G.: Yield gap simulations using ten maize cultivars commonly planted in Northeast China during the past five decades. Agric. For. Meteorol. **205**, 1–10 (2015)
2. Khosla, R.: Zoning in on precision agriculture. Colarado State Univ. Agron. Newsl. **21**, 2–4 (2001)
3. Whelan, B.M., McBratney, A.B., Boydell, B.C.: The impact of precision agriculture. In: Proceedings of the ABARE Outlook Conference, 'The Future of Cropping in NW NSW', Moree, UK, vol. 5 (1997)
4. Jia, L., Yu, Z., Li, F., Gnyp, M., Koppe, W., Bareth, G., Miao, Y., Chen, X., Zhang, F.: Nitrogen status estimation of winter wheat by using an IKONOS Satellite Image in the North China Plain. In: Li, D., Chen, Y. (eds.) Computer and Computing Technologies in Agriculture V. CCTA 2011. IFIP Advances in Information and Communication Technology, vol. 369. Springer, Heidelberg (2011)
5. Stafford, J.V.: Implementing precision agriculture in the 21st century. J. Agric. Eng. Res. **76**, 267–275 (2000)
6. Lamb, D.W., Brown, R.B.: Remote sensing and mapping of weeds in crops. J. Agric. Eng. Res. **78**, 117–125 (2001)
7. Wójtowicz, M., Wójtowicz, A., Piekarczyk, J.: Application of remote sensing methods in agriculture. Commun. Biometry. Crop. Sci. **11**, 31–50 (2016)
8. Kassim, J.M., Abdullah, R.: Advisory system architecture in agricultural environment to support decision making process. In: 2nd International Conference on Digital Information and Communication Technology and its Applications, DICTAP, pp. 453–456 (2012)
9. Bhimanpallewar, R.N., Khade, B.S.: Study of guidelines for agriculture production. ICST-2K14 SBPCOE (2014). ISBN 978-81-928673-0-4
10. Palsaniya, D.R., Rai, S.K., Sharma, P., Satyapriya, Ghosh, P.K.: Natural resource conservation through weather-based agro-advisor. Curr. Sci. **111**, 256 (2016)
11. Pavana, W., Fraissea, C.W., Peresb, N.A.: Development of a web-based disease forecasting system for strawberries. Comput. Electron. Agric. **75**, 169–175 (2011)
12. Olatinwo, R.O., Thara, T.V.: Predicting favorable conditions for early leaf spot of peanut using output from the Weather Research and Forecasting (WRF) model. Int. J. Biometeorol. **56**, 259–268 (2012)

13. Penning de Vries, F.W.T., Jansen, D.M., Ten berge, H.F.N., Bakema, A.: Simulation of ecophysiological processes of growth in several annual crops. IRRi Los Bonos and Pudoc., Wageningen (1989)
14. Oteng-Darko, P., Yeboah, S., Addy, S.N.T., Amponsah, S., Owusu Danquah, E.: Crop modeling: a tool for agricultural research – a review. E3 J. Agric. Res. Dev. 2, 001–006 (2013)
15. Mithra, V.S.S., Somasundaram, K.: A model to simulate sweet potato growth. World. Appl. Sci. J. 4, 568–577 (2008)
16. Kusunose, Y., Mahmood, R.: Imperfect forecasts and decision making in agriculture. Agric. Sys. 146, 103–110 (2016)
17. Tilman, D., Cassman, K.G., Matson, P.A., Naylor, R., Polasky, S.: Agricultural sustainability and intensive production practices. Nature 418, 671–677 (2002)

A Method to Improve the Performance of Raster Selection Based on a User-Defined Condition: An Example of Application for Agri-environmental Data

Driss En-Nejjary[1,2(✉)], François Pinet[2], and Myoung-Ah Kang[1]

[1] LIMOS, University Clermont-Auvergne, Campus Universitaire des Cézeaux, 1 rue de la Chebarde, TSA 60125, CS 60026, 63178 Aubiere Cedex, France
driss.en-nejjary@irstea.fr, kang@isima.fr
[2] Irstea, TSCF, Research Unit "Technologies and Information Systems for Agricultural System", Clermont-Ferrand Centre, 9 Avenue Blaise Pascal, CS 20085, 63178 Aubière, France
francois.pinet@irstea.fr

Abstract. More and more environmental and agricultural data are now acquired with a high precision and temporal frequency. These data are often represented in the form of rasters and are useful for agricultural activities or climate change analyses. In this paper, we propose a new method to process very large raster. We present a new technique to improve the execution time of the selection and calculation of data summaries (e.g., the average temperature for a region) on a temporal sequence of rasters. We illustrate the use of our approach on the case of temperature data, which is important information both for agriculture and for climate change analyses. We have generated several data sets in order to analyze the influence of the different value properties on the process performance. One of our final goals is to provide information about the value conditions in which the proposed processing should be used.

Keywords: Agri-environmental data · Raster selection · Data processing

1 Introduction

The volume of environmental data becomes very important. More and more environmental and agricultural data are now acquired automatically at high precision and temporal frequencies [1–6]. Numerous data sources are available in different information systems [7] and can be accessed through the Web [8–11]. Several of these data sets are useful for agricultural activities or climate change analyses. They can be related to weather, sensor measurements, soil condition, etc. These types of information can be used in agriculture for example for recommendation on the use of agricultural inputs (water, phytosanitary treatments, etc.) or for crop management in order to optimize and reduce the use of agro-equipments, and consequently the negative impact on climate. These data can also be utilized to analyze the links between different agricultural activities (livestock, crops, etc.) and the climate change, at a large spatial and temporal scale.

© Springer Nature Switzerland AG 2019
J. C. Corrales et al. (Eds.): AACC 2018, AISC 893, pp. 190–201, 2019.
https://doi.org/10.1007/978-3-030-04447-3_13

Several of these large data sources are represented in the form of rasters, e.g., geo-referenced regular grids [12, 13]. This type of 2-dimmensional grids constitutes a traditional geographical data format. In geographical information systems, a raster is a 2-D matrix of cells. A measurement (which is very often a numeric value) is stored in each raster cell to represent the geo-referenced value of environmental phenomena: temperature, soil moisture, CO_2 measurements, rain precipitation, etc. This type of data can also be produced by simulation.

Dedicated methods are needed to manage the huge volume of rasters produced over time. It is important to propose specific techniques to optimize the analysis and the processing of such data sources.

In this paper, we propose a new method to analyze large sets of rasters. Our goal is to propose a new technique to improve the execution time of the selection and calculation of data summaries (e.g., the average temperature for a region) on a temporal sequence of rasters. We illustrate the use of our approach on the case of temperature data, which is crucial information both for agriculture and for climate change analyses.

Section 2 describes more precisely the raster data process discussed in this paper and highlights its interest. Section 3 introduces our heuristic to improve the performance in terms of processing time. Sections 4 and 5 show some first experiments of this method on simulated data. Section 6 presents conclusion and future work.

2 Raster Data Process

The raster data process used in this paper consists in three main steps (shown in Fig. 1). The process is based on a large sequence of rasters, representing the evolution of an environmental pheromone over time. In Fig. 1, the different values of the raster cells are represented by colors. In the step (a), the user chooses a period of interest. More precisely, he/she selects a temporal raster (sub)sequence of interest in the large sequence of rasters. In the step (b), the user defines the geographical region to analyze in the sequence of rasters selected in step (a). This geographical region to analyze is the same for all these rasters. In the step (c), the system automatically selects every raster that satisfies a user-defined condition. Figure 2 provides a UML representation of the different object types.

We illustrate this process on an example. A user would like to analyze a sequence of rasters representing the evolution of temperatures. He/she wants to determine the set of rasters having low temperatures in order to:

- study more precisely these cases and their possible local causes. It is a typical case of climate change analysis.
- or analyze the impact of these temperature on crops in agriculture in the context of farm decision support.

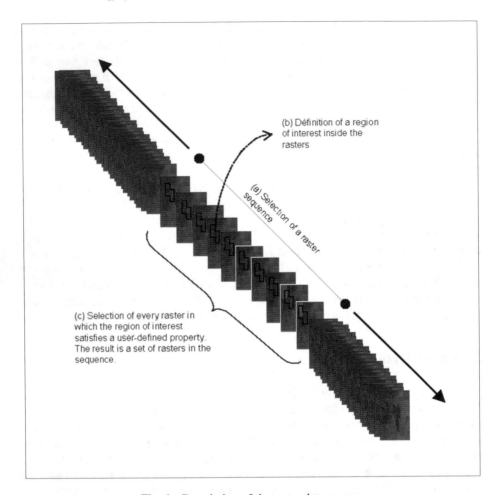

(b) Définition of a region of interest inside the rasters

(a) Selection of a raster sequence

(c) Selection of every raster in which the region of interest satisfies a user-defined property. The result is a set of rasters in the sequence.

Fig. 1. Description of the raster data process.

First, he/she manually chooses the period to be analyzed in the whole sequence (step (a)). Second, he/she manually chooses a geographical region of interest for his/her study (step (b)). Third, in the step (c), the user would like to automatically select every raster in which the average temperature of the region of interest is lower than a user-defined threshold (e.g., $\leq 10\,°C$). Consequently, the result is the set of the rasters that satisfy this condition.

In our scenario, the steps (a) and (b) imply a manual user choice. The step (c) can be automated, e.g., the calculation of the average temperature for every raster and the selection of the temperature $\leq 10\,C°$. A naive algorithm for the step (c) is shown in Fig. 3:

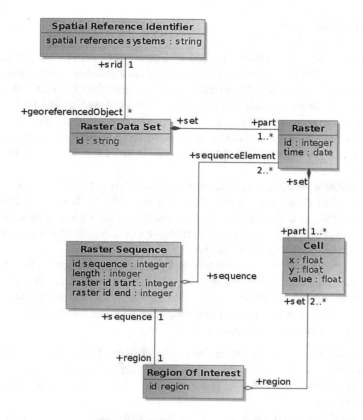

Fig. 2. Raster data set modelling.

S := the whole sequence of rasters
A := the (sub)sequence of rasters chosen in S by the user
b := the region of interest chosen by the user/
$Result$:= { }

for every raster Ri in A
{
    avg := the cell average for the region b in Ri
    if $avg < 10$ then $Result$:= $Result \cup \{ Ri \}$
}

Fig. 3. Naive algorithm for step (c)

3 Raster Data Process

In the present paper, we propose a method to improve the performance for the step (c). The intuition behind this algorithm is to try to reject a raster that does not satisfy the user-condition (i.e., the cell value average must be lesser then the user's threshold) as soon as possible to avoid useless computation. The proposed technique can improve the computation when the user's threshold is low (compared to the raster cell values). In this technique, the cell must contain only numerical positive value – consequently, a uniform translation or normalization must be used if the rasters do not comply with this constraint.

The calculation of the average is computed for each raster (in the region of interest). The average computation consists in calculating the sum of cell values for each raster. In the new version of our algorithm, we stop the sum computation as soon as possible, when we are sure that this sum becomes superior to the threshold value multiplied by the cell number of the region of interest.

We also propose to sort the cell values in the region of interest in a descending order, for the average computation. In that case, the threshold is reached faster for the rasters that do not satisfy the condition. Unfortunately, the time complexity of a sort, i.e., $O(n \log n)$ for a quick sort, is higher than the sum computation, i.e., $O(n)$. Consequently, we propose the following stages:

(1) We propose to sort the value of the region of interest only for some rasters, e.g., compute a sort every 200 rasters, in sorting the cell values of the region of interest only for the rasters R_i, R_{i+200}, R_{i+400}, etc. Each one of these sorts produces a cell ordering.

(2) We propose to use the cell ordering of the sorted rasters, for computing the sums for the other rasters. For example, the sort in R_i produces a cell ordering. This cell ordering will be used for computing the sum for each raster from R_i to R_{i+199}. The cell ordering determined by the sort of R_{i+200} will be used for each raster from R_{i+200} to R_{i+399}, etc.

The intuition behind this method is that in many phenomena the spatial distribution of values evaluates rather slowly over time. In the case of temperature rasters produced every 5 min, the highest values will often be on the same geographical part of the rasters for several tens of minutes or several hours. The frequency of the sort computation can be adapted to the nature of the data (e.g., sorting every 10 rasters, 50 rasters, 100 rasters, 200 rasters, etc.). This new version of the algorithm is shown in Fig. 4. In Fig. 4, *Ord* is an array that corresponds to a mapping: *Ord(1)* is equal to the cell number (#) in *b* that has the highest value; *Ord(m)* is equal to the cell number (#) in *b* that has the lowest value.

Several constraints must be satisfied in order to guarantee that this algorithm provides better performances in terms of execution time, for example, a low user-defined threshold or a spatial distribution of cell values sufficiently large in every raster to justify the interest of the sorting operation. These aspects and several proposed improvements are discussed in the last section of the paper.

Improved algorithm for step (c):

S := the whole sequence of rasters
A := the (sub)sequence of rasters chosen in S by the user
b := the region of interest chosen by the user
m := the number of cells in b
$begin$:= the number (#) of the first raster in S
end := the number (#) of the last raster in S
it := the interleave between two consecutive rasters on which a sort is calculated
th := the user-defined threshold

$Result$:= { }
$maxsum$:= $th * m$

for i := $begin$ to end step it
{
    Sort the cell values of the region b in R_i in descending order
    and produce the corresponding cell ordering Ord

    for j := i to $i+it$
    {
      if $j > end$ then { process completed ; stop } else
      {
        su := 0
        $reject$:= false
        for k := 1 to m
        {
          v := the value in the $cell_{Ord(k)}$ in R_j
          su := $su + v$
          if $su > maxsum$ then { $reject$:= true ; break}
        }
        if $reject$ is false then $Result$:= $Result \cup$ { Rj }
      }
    }
}

Fig. 4. Improved algorithm for step (c)

4 Experiment Description

To evaluate the performance of our improved algorithm of raster selections, we ran our experiments on the Intel(R) Core(TM) i5-5350U CPU at 1.8 GHz. Concerning the data, the experiments are conducted using a public data set.

We used the public dataset provided by the US National Oceanic and Atmospheric Administration [14]. This dataset provides a large amount of climate and historical weather data, such as: air temperature, humidity, precipitation, etc. available in different temporal acquisition frequencies: monthly, daily, hourly and sub-hourly (5-minute). These data are produced from many weather stations in the USA and elsewhere. In our work we have used the daily frequency of data acquisition for four years 2014, 2015, 2016 and 2017 of the Harrison station in USA. Each day, we have the min, max and the average of the temperature of this latter. We have chosen the daily temperature to have a significant difference between the minimum temperature and the maximum temperature for the station, the thing that we need in our algorithm, which is not the case for the hourly data for instance. In order to obtain a large and significant data set, we have generated (i.e., simulated) rasters from the station data.

To build our dataset, we have simulated rasters for the local studied region. We assumed that the temperature of a region has a Gaussian distribution. The normal distribution has two characteristics: the mean and the standard deviation. The mean is included in the initial dataset, and we have estimated the standard deviation ST using the min and the max values (also provided by the initial dataset). The simple method to simulate the standard deviations ST is the range rule of thumb [15]. Here is the calculation of the approximation of ST:

$$ST \approx \frac{max - min}{4}$$

In practice, the estimation of ST using the range rule of thumb is not sufficient when the n is extremely small or large. The authors of [15] have improved this estimation in order to deal with this problem:

$$ST \approx \begin{cases} \frac{1}{\sqrt{12}}\left[(max - min)^2 + \frac{(max - 2m + min)^2}{4}\right]^{1/2} & n < 15 \\ \frac{max - min}{4} & 15 < n < 70 \\ \frac{max - min}{6} & n > 70 \end{cases}$$

In our experiment, we have simulated a large data set. Consequently, we used the third case for the estimation of our standard deviation ($n > 70$). Thus, we had the required parameters to produce raster data from the initial data using a normal distribution.

We have created 3 data sets having 3 different raster sizes; each data set contains 1420 rasters. We have applied a translation on the cell values in order to avoid negative value; we have added a constant (50) to each cell value. In our tests, the user-defined region of interest is the whole raster. In the produced data sets, we have one raster every day for four years. The tests have been applied on all these rasters – these rasters constitute the sequence A of rasters to analyze.

5 Results

In this section, we show the results of our experiments performed on the generated data. Different raster sizes have been tested. For each experiment, we test the naive algorithm and the improved version on the same data set. In our experiments, we also evaluate the impact of the main parameters on the execution time of our algorithm, for instance, the threshold and the interleave between the sorted rasters. To do this, we have chosen different thresholds and interleaves and run our algorithm using these different value parameters. Concerning the sort algorithm, we used a quick sort.

5.1 The Impact of the Threshold on the Performance

Tables 1, 2 and 3 compare the computing time for the naive and the improved algorithm for the three data sets for different threshold values.

Table 1. Dataset 1: Size of raster = 100 × 100, contains 1420, Interleave = 73

	Threshold $th = 30$	Threshold $th = 40$	Threshold $th = 45$	Threshold $th = 46$	Threshold $th = 50$
Naive algorithm	8.5(s)	13.17(s)	12.30(s)	14(s)	14.91(s)
Improved algorithm	7.9(s)	10.1(s)	11.9(s)	12.9(s)	13(s)

Table 2. Dataset 2: Size of raster = 200 × 200, contains 1420, Interleave = 73

	Threshold $th = 39$	Threshold $th = 40$	Threshold $th = 41$	Threshold $th = 42$	Threshold $th = 50$
Naive algorithm	44.57(s)	45.30(s)	47.80(s)	49.4(s)	53.76(s)
Improved algorithm	40.5(s)	42.28(s)	43.04(s)	44.47(s)	50.39(s)

Table 3. Dataset 3: Size of raster = 240 × 240, contains 1420, Interleave = 73

	Threshold $th = 30$	Threshold $th = 40$	Threshold $th = 42$	Threshold $th = 50$	Threshold $th = 70$
Naive algorithm	36.81(s)	46.63(s)	48.53(s)	56.83(s)	67.03(s)
Improved algorithm	32.57 (s)	44.57(s)	46.78(s)	53.55(s)	62.40(s)

Table 1 shows that the improved algorithm is faster than the naive one, especially when the threshold is not too low and not too high. The best performance is with $th = 40$; our algorithm I is faster than the naive one with 3.07 s less for time execution. Whereas when the threshold is smaller, we obtain less performance (the case of $th = 30$).

In Table 2, our algorithm is faster than the naive one with 5 s less for execution time ($th = 42$).

As we can see in the Table 3, our algorithm is still faster than the naive one. More precisely, our algorithm is always faster than the naïve one, whatever the value of the threshold. The user-defined threshold value has a direct impact on the performance of the improved algorithm.

5.2   The Impact of the Interleave Size on the Performance

The interleave value between the sorted rasters is important. It has also an impact on the performance of our algorithm. Choosing a low interleave implies sorting more rasters, which decreases the performance. In the other hand, choosing large interleave means sorting less rasters which is good for the performance, but in the same time, many rasters that are in the same interleave will not follow the same behavior as the sorted raster.

In Table 4 we show how the interleave size influences the performance of our algorithm on the data set 1. As an example, we have tested three interleave sizes. As we can see in Table 4, the best performance is obtained by choosing the size 73. The choice of the interleave value depends on the nature of data and the frequency of its production.

Table 4. The impact of the interleave on the performance (Data set 1), Threshold = 40

	Interleave =10	Interleave =20	Interleave =73
Naive algorithm	13.17(s)	13.17(s)	13.17(s) (s)
Improved algorithm	11.27(s)	10.53(s)	10.4

Our algorithm shows interesting potential, it should be improved by using other faster sorting algorithms and also using raster data sets with significant variation of data in the same raster.

6   Conclusion and Discussion

In literature, different proposals have been implemented for raster processing, but no previous work had proposed our approach based on a value sort for a conditional selection of rasters. The processing proposed in this paper is rather specific and is in the field of the conditional selection, based on a maximal threshold, in a large raster data set. It corresponds to a concrete and useful operation for processing the set of rasters especially in the context of global climate change and agri-environment. Consequently, it was important to propose and develop new techniques to improve the computation.

We show in this paper, an idea for improved processing and the first associated tests. We have generated several data sets in order to analyze the influence of the different value properties on the process performance. We have to continue our tests to highlight in which cases our approach is suitable. For that, we have to analyze more precisely the impacts of the data, the interleave, the threshold, the number of selected rasters, etc. on the performance of the process. Our goal is to provide to computer scientists, information about the conditions in which the proposed improvement is interesting and should be used. Thanks to this information, we will produce recommendations about the different types of agri-environmental data suitable for this technique. Our proposal can also be directly applied on large vector of geo-referenced sensor data. We also plan to test our method on a real data set produced by sensor acquisition in real farms (soil and air moisture, temperature, etc.).

Here, we emphasize two main perspectives for the future improvement of our method

(1) The use of the General-Purpose logic on Graphics Processing Unit (GPGPU) technology can be experimented in order to provide a parallel processing of the computation. Different tests have been made for the optimization of operations on rasters using GPGPU by other authors [16–19]. Currently, there is no implementation proposal by GPGPU for our improved process. Our process can be massively parallelized, as several parts of the procedure are independent and can be run in parallel. For example, several rasters can be processed in parallel.

(2) It is useless to pre-calculate and store the average of each raster in a persistent manner. The user chooses a sub-part of the raster (i.e., the geographical region of interest) and this sub-part is not known in advance. This region of interest can also be different for each new user query. Nevertheless, we could decompose each raster in several partitions (e.g., regular partitions) and pre-calculate and store persistently the average of the cell values for each partition. In this case, we could use these averages as indicators to determine if our improved process should be used or not. For example, if the user-defined geographical region of interest is spatially included in a partition p, the system could decide to automatically run our improved method if the partition average (of the first raster in the sequence for example) is greater than the user-defined threshold. Different other parameters could be used to determine if it is useful to run our improved method.

To conclude, in our opinion, the research work presented in this paper opens the way to other interesting contributions that can be of interest in the field of agri-environment, especially for sensing data produced by remote sensing, agricultural internet of things and smart farming [20–26].

Acknowledgement. This work was funded by grants from the French program «investissement d'avenir» managed by the Agence Nationale de la Recherche of the French government (ANR), the European Commission (Auvergne FEDER funds) and «Région Auvergne» in the framework of LabEx IMobS 3 (ANR-10- LABX-16-01). We also acknowledge the support received from the Agence Nationale de la Recherche of the French government through the program «investissement d'avenir» 16-IDEX-0001 CAP 20-25 on the general topic of this paper.

References

1. Sawant, S., Durbha, S.S., Jagarlapudi, A.: Interoperable agro-meteorological observation and analysis platform for precision agriculture: A case study in citrus crop water requirement estimation. Comput. Electron. Agric. **138**, 175–187 (2017)
2. Lee, W.S., Ehsani, R.: Sensing systems for precision agriculture in Florida. Comput. Electron. Agric. **112**, 2–9 (2015)
3. Ahmedi, F., Ahmedi, L., O'Flynn, B., Kurti, A., Tahirsylaj, S., Bytyçi, E., Sejdiu, B., Salihu, A.: InWaterSense: An intelligent wireless sensor network for monitoring surface water quality to a river in Kosovo. Int. J. Agric. Environ. Inf. Syst. **9**, 39–61 (2018)
4. Devadevan, V., Sankaranarayanan, S.: Forest fire information system using wireless sensor network. Int. J. Agric. Environ. Inf. Syst. **8**, 52–67 (2017)
5. Kang, M.A., Pinet, F., Schneider, M., Chanet, J.P., Vigier, F.: How to design geographic database? Specific UML profile and spatial OCL applied to wireless Ad Hoc networks. In: 7th Conference on Geographic Information Science (AGILE 2004), Heraklion, GRC, April 29–May 1 2004, pp. 289–299 (2004)
6. Pinet, F.: Entity-relationship and object-oriented formalisms for modeling spatial environmental data. Environ. Model. Software **30**, 80–91 (2012)
7. Pinet, F., Miralles, A., Papajorgji, P.: Modeling: a central activity for designing and implementing flexible agricultural and environmental information systems. Int. J. Agric. Environ. Syst. **1** (2010)
8. Machwitz, M., Hass, E., Junk, J., Udelhoven, T., Schlerf, M.: CropGIS – a web application for the spatial and temporal visualization of past, present and future crop biomass development. Comput. Electron. Agric. (2018)
9. Roussey, C., Bernard, S., Pinet, F., Reboud, X., Cellier, V., Sivadon, I., Simonneau, D., Bourigault, A.-L.: A methodology for the publication of agricultural alert bulletins as LOD. Comput. Electron. Agric. **142**, 632–650 (2017)
10. Ramar, K., Gurunathan, G.: Semantic web based agricultural information integration. Int. J. Agric. Environ. Inf. Syst. **8**, 39–51 (2017)
11. Rabindra, B.: CloudGanga: cloud computing based SDI model for ganga river basin management in India. Int. J. Agric. Environ. Inf. Syst. **8**, 54–71 (2017)
12. Laurini, R., Thompson, D.: Fundamentals of spatial information systems (1992)
13. Kang, M.-A., Zaamoune, M., Pinet, F., Bimonte, S., Beaune, P.: Performance optimization of grid aggregation in spatial data warehouses. Int. J. Digital Earth **8**, 970–988 (2015)
14. Diamond, H.J., Karl, T.R., Palecki, M.A., Baker, C.B., Bell, J.E., Leeper, R.D., Easterling, D.R., Lawrimore, J.H., Meyers, T.P., Helfert, M.R., Goodge, G., Thorne, P.W.: U.S. Climate Reference Network after one decade of operations: status and assessment. Bull. Amer. Meteor. Soc. **94**, 489–498 (2013). https://doi.org/10.1175/BAMS-D-12-00170.1
15. Hozo, S.P., Djulbegovic, B., Hozo, I.: Estimating the mean and variance from the median, range, and the size of a sample. BMC Med. Res. Methodol. **5**, 13 (2005). https://doi.org/10.1186/1471-2288-5-13
16. Melesse, A.M., Weng, Q., Thenkabail, P.S., Senay, G.B.: Remote sensing sensors and applications in environmental resources mapping and modelling. Sensors **7**, 3209–3241 (2007)
17. Zhang, J., You, S., Gruenwald, L.: Parallel online spatial and temporal aggregations on multi-core CPUs and many-core GPUs. Inf. Syst. **44**, 134–154 (2014)
18. Prasad, S.K., McDermott, M., Puri, S., Shah, D., Aghajarian, D., Shekhar, S., Zhou, X.: A vision for GPU-accelerated parallel computation on geo-spatial datasets. SIGSPATIAL Special **6**, 19–26 (2015)

19. Simion, B., Ray, S., Brown, A.D.: Speeding up Spatial Database Query Execution using GPUs. Procedia Comput. Sci. **9**, 1870–1879 (2012)
20. Data management and analytics in interdisciplinary research. Comput. Electron. Agric. **145** 130–141 (2018)
21. Lehmann, R.J., Reiche, R., Schiefer, G.: Future internet and the agri-food sector: state-of-the-art in literature and research. Comput. Electron. Agric. **89**, 158–174 (2012)
22. O'Grady, M.J., O'Hare, G.M.P.: Modelling the smart farm. Inf. Process. Agric. **4**, 179–187 (2017)
23. Severino, G., D'Urso, G., Scarfato, M., Toraldo, G.: The IoT as a tool to combine the scheduling of the irrigation with the geostatistics of the soils. Future Gener. Comput. Syst. **82**, 268–273 (2018)
24. Talavera, J.M., Tobón, L.E., Gómez, J.A., Culman, M.A., Aranda, J.M., Parra, D.T., Quiroz, L.A., Hoyos, A., Garreta, L.E.: Review of IoT applications in agro-industrial and environmental fields. Comput. Electron. Agric. **142**, 283–297 (2017)
25. Tzounis, A., Katsoulas, N., Bartzanas, T., Kittas, C.: Internet of Things in agriculture, recent advances and future challenges. Biosyst. Eng. **164**, 31–48 (2017)
26. Wolfert, S., Ge, L., Verdouw, C., Bogaardt, M.-J.: Big data in smart farming – a review. Agric. Syst. **153**, 69–80 (2017)

Coffee Crops Variables Monitoring: A Case of Study in Ecuadorian Andes

Juan Abad[1(✉)], Juan Farez[1(✉)], Paúl Chasi[1], Juan Carlos Guillermo[1(✉)], Andrea García-Cedeño[1(✉)], Roger Clotet[2(✉)], and Mónica Huerta[1(✉)]

[1] GITEL, Telecommunications and Telematics Research Group, Universidad Politécnica Salesiana, Cuenca, Ecuador
{jabadb,jfarez}@est.ups.edu.ec
{pchasi,jguillermo,agarciac,mhuerta}@ups.edu.ec
[2] Networks and Applied Telematics Group, Universidad Simón Bolívar, Caracas, Venezuela
clotet@usb.ve

Abstract. Coffee bean is one of the most cultivated products worldwide. Ecuadorian Andes have good conditions for the production of coffee, but many farmers have small plantations which difficult utilization of technology to improve crops productivity. Few studies focused on analyzing climatic variables that affect the optimum development of the coffee plant. We have developed a low cost a system to monitor environmental and soil parameters in coffee plantations in real time, in addition of keeping historical records of the measurements, is designed, developed and implemented to facilitate farmers management of crops. Also the system compare sensed values with optimum range for each of them and alert farmer if some action is need using a user-friendly interface.

Keywords: Monitoring · WSN · Coffee crops

1 Introduction

The Food and Agriculture Organization of the United Nations (FAO) indicates that the number of people suffering from hunger in the world continues to increase, reaching 821 million in 2017. Hunger has grown in the last three years, returning to the levels of ten years ago. This reversal in progress sends a clear alert that more must be done and urgently if the Sustainable Development Goal of Zero Hunger is to be achieved by 2030. The situation is getting worse in South America and in most regions of Africa. The climatic variability and meteorological phenomena are the main causes of the reduction of agricultural activity in these regions. Another cause of the reduction of agricultural production is the lack of automation and monitoring [1]. There are several research studies that show that the monitoring of climatic variables optimizes the production of various crops, as is the case of tomato [13,25,32,33], habanero pepper [9], cassava

J. C. Corrales et al. (Eds.): AACC 2018, AISC 893, pp. 202–217, 2019.
https://doi.org/10.1007/978-3-030-04447-3_14

crops [8], oil palm [10,21,29], flowers [12,14], cacao [18,19], coffee [22], among others.

Coffee bean is among the 21 most cultivated products worldwide as the produced beverage leads to high consumption rates, generating the birth of hundreds of national producers in various coffee growing countries. Ecuador has a political-administrative division of 24 provinces, the coffee area, estimated by the National Coffee Council (COFENAC), corresponds to 52,538 hectares, established in the main central and southern areas of Ecuador [20].

Given the geographical conditions of Ecuador, coffee is grown throughout regions: coast, mountains, Amazon forest and Galápagos. This is one of the few countries in which all types of coffee are produced: Robusta (*Coffea canephora*) and Arabic (*Coffea arabica*) derivatives, and addition to bean production, it also exports industrialized coffee [7]. These two varieties of coffee have a similar behavior in the cultivation area, which is located mainly in the southern zone of Ecuador with an altitude between 700–1800 m above sea level, making it more suitable for Arabica coffee, this type is the one that covers the largest area of crops, generating a production of 115,000 quintals according to the ecuadorian corporation of coffee growers (CORECAF) [28].

The strategy of maximizing coffee production will not only substantially increase yields, but also reduces production costs and heightens resistance to pesticides, which expose the health of crop personnel and the surrounding environment. A study by the Food Web in coffee plantations shows a reduction in their lifespan, when growing conditions are affected by several variables, including the level of light, associated humidity and pest development [27]. Different levels of shade provide the basis for the identification of optimal variable conditions that minimize the entire pest complex and maximize the effects of beneficial microflora and fauna that act against it, these conditions of optimal variables for plant development differ with the climate, altitude and soil [5,24]. Coffee plantations in the country do not possess sensor networks to monitor agricultural variables, due to the incursion of telecommunications, monitoring is carried out manually, based on the experience of a person who describes whether the climate is favorable or not. In order to change the way of measuring the variables involved in cultivation, devices will be designed which allow the measurement of climatic variables in a coffee plantation in an automated way [2].

Some studies have focused on optimizing coffee production based on different techniques such as: image analysis of grain and branches [6,31], grain classifiers using artificial intelligence [23,30], disease and pest detection [3,4,26], but there are few studies focused on analyzing the climatic variables that affect the optimum development of the coffee plant. In this work, a Wireless Sensor Networks was designed and implemented to monitor climatic variables in a coffee crop.

This document is distributed as follows: background information about coffee production in Ecuador and the principal variables to take into account for its cultivation are explained; in Sect. 2, the methodology summarizing the different stages for this proposal development will be detailed; Sect. 3 is focused on the results; and finally, in Sect. 4, conclusions will be held.

2   Coffee Crops Monitoring

As detailed above, coffee can be grown in different regions throughout Ecuador. In particular for this study a coffee plantation located in Ecuadorian Andes, specifically in the canton of Santa Isabel in Azuay province, has been selected. Figure 1 shows the location. Santa Isabel has a microclimate that is very suitable for coffee crops due their particular temperature range, precipitation, soil or number of sun hours per day. Ministry of Agriculture and Livestock of Ecuador had promoted the recovery of this zone providing farmers with new coffee plants to plant or to substitute old ones. All or almost all farmer's crops are familiar or small plantations making very difficult for them to access to technological tools to monitor their crops due costs.

Fig. 1. Santa Isabel plantations location

2.1   Santa Isabel Plantations

Santa Isabel coffee producers' association (ASOPROCSI) was created in 2010 to help producers to managed their plantations and cooperate between them. Cumbis and Jiménez [11] studied coffee plantations of this sector of Ecuador detecting that productive part is deteriorated, the plantations are old, with production of only 5 to 6 quintals/hectare. To improve production volume per hectare and coffee grain quality its fundamental to monitor some environmental variables allowing farmers to take some actions to improve or counteract adverse conditions, for example irrigation of soil when soil moisture is too low [15].

Considering environmental parameters that affect the development of a coffee plantation, the variables to be monitored are: Temperature, pH, soil moisture, environmental humidity, carbon dioxide (CO_2) and luminosity.

Some studies determine that the optimum values for this variables must be within the minimum and maximum to guarantee an optimal growth and quality. Table 1 shows required range for each parameter [17].

Table 1. Optimum values for coffee cultivation [16].

Coffee cultivation values range					
Altitude (masl)	°C	Humidity	Sunshine hours per year	CO^2	pH
600–1800	18–24	70%–85%	1600–1800	700–900	5–5.5

2.2   Technical Features

The developed system has the following characteristics of wireless communication, management and data access. As shown in Fig. 2.

Power Module: This module is designed based on energy saving principles, the acquisition devices will wake up specifically every hour to measure the climatic conditions of the coffee plantation and then they will return to sleep mode. The total power consumption of the transceiver system is 54 mA, and according to the Ecuador's Solar Atlas for electricity generation purposes, the study area insolation value corresponds to 2.5 KWh/m^2/day. Hence, in based of mentioned values, the capacities of the battery and the photovoltaic panels were dimensioned to achieve the system to have reserved energy when there is no presence of sun, resulting in 2Ah batteries and 6 V 200mA panels.

Every sensor node counts with an integration of batteries and solar panel that makes it energetically self-sustainable since each battery is responsible for supplying the necessary power when no solar energy is present. The photovoltaic power supply process is: 6 V solar panel -> 3.7 V -> Voltage booster 3.7 V to 5 V -> 5 V power supply.

Acquisition Module: The acquisition module consists of five data collection sensors which are connected to an ATmega328P microcontroller found on the Arduino NANO board, this device is the one in charge of processing each of the measurements from the different sensors.

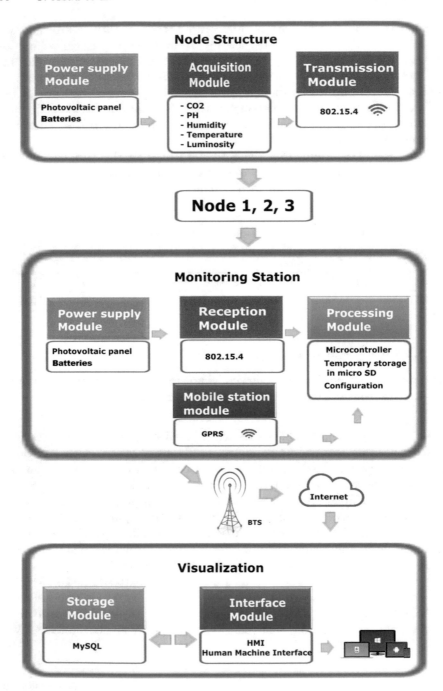

Fig. 2. Block diagram of implemented system

The measurements from the coffee plantation are collected in a resolution of 10 bits to be able to detect level variations in the input signal up to 5 mV, being aware that Xbee module transmitter (tx) bits must be of a length of 8 for each of the measurements, it is necessary to perform a bit shift and therefore split the data frame into 2 bytes by means of a bit masking. Of all the sensed variables, the pH measurement at a given time reaches values of 855, this value is used to demonstrate the fractioning process in Fig. 3.

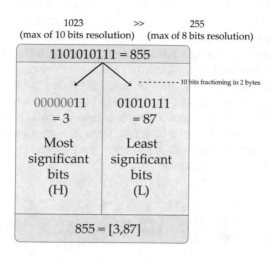

Fig. 3. 10 bits data fractioning to fit 8 bits requirement for Xbee TX

Carrying out this process to each of the climatic variable readings, the frame detailed in Table 2 is formed.

Table 2. Screen formed with the measurements collected by the sensors

Header(bit0-bit15)	Hum		Temp		CO^2		Lux		PH		Checksum
0×7E	bit H	bit L	bit H	bit L	bit H	bit L	bit H	bit L	bit H	bit L	total sum of bits

Transmission and Reception Module: The transmission module consists of an XBee S2C-PRO which allows creating an access point for the mesh sensors network, unifies and interconnects the final devices to send the data packets to the coordinator module. The sending of packets is done through the 2.4 GHz band using the DSSS modulation technique. Each data packet consists of the unique address of the device, followed by the composition of the frame composed of the acquisition of the module.

The receiver module is composed of an XBee configured in coordinator mode, which is responsible for maintaining the network in proper operation. It can not be configured in energy saving mode, it is always sending and receiving data from modules that have the same PAN ID. It is designed to avoid the collision of data if they arrive at the same time and is responsible for queuing them in order of arrival, thus avoiding the loss of data packages.

Processing Module: It is formed by a micro-controller that is embedded in the coordinating node. The micro-controller has the function of processing each of the frames that arrive from the three sensor nodes. The scheme for processing data from the sensor nodes is shown in the Fig. 4, which is described below: Within the sensor network, a recognition of the sector is carried out by means of the MAC of the device, then data is obtained from the humidity, temperature, luminosity and CO_2 sensors, thus forming a data frame together with the payload. Once the complete frame of each sensor node is obtained, it is temporarily stored in a micro-SD in a .txt file, then it is analyzed, structured and processed. Finally, with the complete data-set, we proceed to store permanently in the database of the remote server.

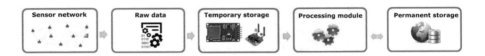

Fig. 4. Information flow

The microprocessor is also in charge of sending configuration commands to the sensor nodes and updating the sampling time. To configure the device, the frame is armed with the identification of the recipient element along with the sampling time, the query of the status of the nodes and the request for current data.

Mobile Station Module: GPRS is a mobile device that works at 850/900/1800/1900 Mhz and can be used for text messages, calls, Internet access to transport data to the server from anywhere with 9.6 kbps of speed. It allows to send the data of the climatic variables towards the remote server, for this it is necessary to configure through AT commands: the GPRS mobile network (internet access, HTTP protocol for transport), configure the reading of text messages, check internet connection, count on the connection to the GPRS network, assign a user name and password, etc. Once the GPRS module is configured, the micro SD is accessed to read the five parameters and at the same time the link to the server is armed. The GPRS module allows you to go from a private network to a public network, thanks to the fact that it is connected to the Internet, as current mobile devices do.

Data Base Module: It allows to execute a script programmed in Python to store all the parameters of the sensor nodes sent from the coordinator node in the MySQL database. The storage of the parameters is done through the following nomenclature: T = 400, H = 700, PH = 8.4, C = 200 and L = 900. In addition to the monitored parameters are stored the states of each sensor node (on or off), as well as the current Date Time. Users can retrieve these parameters via GET requests through a URL, returning the initial letter of each parameter with its respective value. Before displaying the parameters internally these are transformed to the corresponding unit. For example the temperature value 1850 is divided by 100 to obtain 18.50 °C, the same is done for PH and brightness. The humidity value is converted to percentage and the CO_2 value is converted into units parts per million (ppm). Depending on the sector, a certain script is executed.

3 Results

The area of the crop is divided into three sectors called A, B and C. Sector A has plants in growth stage, sector B counts with plants in full development, and sector C has coffee plants that need to be transplanted. In each crop area a node is placed and identified with the same name as the sector where it is located. Figure 5 shows the sectors where nodes were placed and the distances between them. Each of them has a weather protection ensuring that the internal components are not affected. Additionally, the sensor cables are coated to protect

Fig. 5. Coffee monitoring WSN deployment location

them from external agents such as animal gnawing or climatic factors. Each node have photovoltaic panels to energize, panel angle and inclination are set considering region, latitude and the seasons.

The interface with the user was made using open source software using client–server model. Client can access using any web browser to server to retrieve nodes information, of course once is validated in the system. The start window shows the last stored values of each sector with its respective date and shows the map where the project was developed as show Fig. 6. The lower part shows the alarms generated in the plantation when climatic parameters exceeds the optimum thresholds establish in this case for coffee plantation.

3.1   Graphic User Interface

According to the design procedure, network nodes and their deployment, it is possible to obtain the measurements of the coffee plantation, which let us validate the network of sensors and the correct system operation. New cycle of readings always start at 6:00am everyday. Acquired data allow analysis and visualization of behavior of following parameters: Humidity, pH, CO_2, temperature and luminosity as show in Fig. 6.

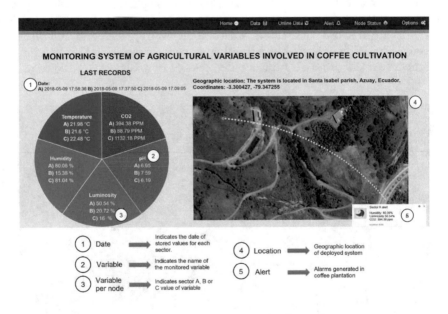

Fig. 6. System homepage visualization

Environmental and soil parameters values can be consulted in real time for each sector (A, B or C). System feedback user with green color (in optimum range) or red color (out of optimum range) for all parameters, allowing the farmer to detect quickly that needs attention and allow him to take corrective measures (as shown in Fig. 7).

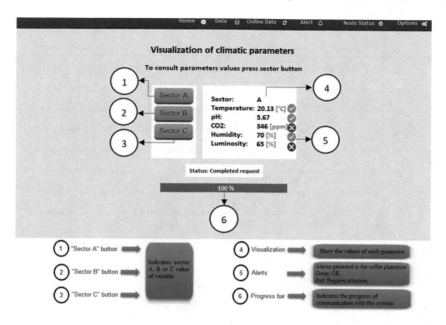

Fig. 7. Visualization of climatic parameters

3.2   Results of Monitored Parameters

Measurements were made for two days of all parameters in each sector staring at 6:00am. Figure 8 shows environmental humidity in sector C remains relatively constant, this were two days without rains and the farmers did not realize that it was necessary to irrigate to increase humidity.

In case of pH, data obtained in sector B show that this soil is alkaline but not enough to be in optimum range (5 to 5.5). Sensor detect a maximum pH of 6.2 and a minimum of 5.45, but normally it maintains upper of 5.5, as shown in Fig. 9. Needing the intervention of the farmer, using for example lime, to slightly increase the alkalinity of the soil.

CO_2 values for three day monitoring in sector C shows a behavior related to the time of day, repeating a similar pattern every day as show Fig. 10. Coffee plants perform photosynthesis during the day and at night this chemical process is reduced.

Temperature monitoring in sector B show a minimum value of 15.5 °C in the early morning hours, increasing gradually to a maximum of 34.8 °C after midday, as shown in Fig. 11. This is the typical temperature pattern for the Andean climate.

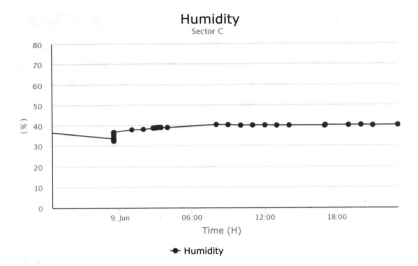

Fig. 8. Humidity values.

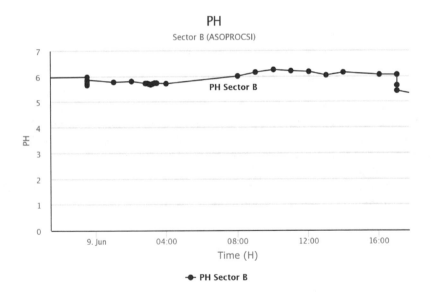

Fig. 9. pH values.

As a typical Andean climate pattern, during monitoring of sector C we can observe an intense luminosity during daytime, as shown in Fig. 12. Although some sunshine hours are good for coffee crops, in this climate farmers experience recommend providing some shade to avoid overexposure.

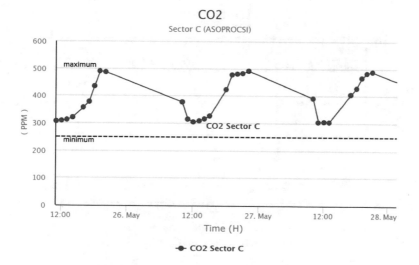

Fig. 10. CO_2 values.

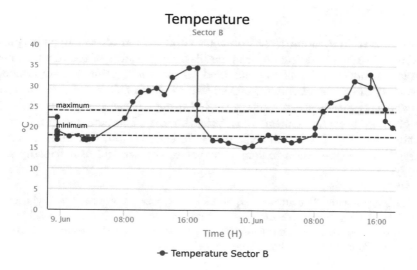

Fig. 11. Temperature values.

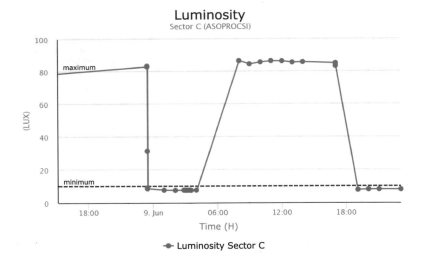

Fig. 12. Luminosity value.

4   Conclusions

A low cost system to monitor environmental and soil parameters in coffee plantations in real time, in addition of keeping historical records of the measurements,
is designed, developed and implemented to facilitate farmers management of
crops. Also the system compare sensed values with optimum range for each of
them and alert farmer if some action is need using a user-friendly interface. Each
node of the system can monitor: CO_2, pH, Humidity, Temperature and Luminosity (sunshine). Equipped with solar panels to energize and transmission module
to transfer recollected data to central server to storage and user-friendly access
via web.

Acknowledgement. The authors would like to express their gratitude to the support
of the Thematic Network: RiegoNets (CYTED project 514RT0486), as well to PLA-
TANO project by Telecommunications and Telematics Research Group (GITEL) from
Universidad Politécnica Salesiana, Cuenca - Ecuador.

References

1. Global hunger continues to rise, new un report says. Technical report, Food and
   Agriculture Organization of the United Nations (FAO) (2018). Accessed 17 Sept
   2018
2. Abad Alameda, A.S.: Diseño e implementación de un sistema de seguimiento de
   parámetros ambientales en plantaciones de café (2017)

3. Ahmed, R.Z., Biradar, R.C.: Data aggregation for pest identification in coffee plantations using wsn: a hybrid model. In: 2015 International Conference on Computing and Network Communications (CoCoNet), pp. 139–146. IEEE (2015)
4. Ahmed, R.Z., Biradar, R.C.: Energy aware routing in WSN for pest detection in coffee plantation. In: 2016 International Conference on Advances in Computing, Communications and Informatics (ICACCI), pp. 398–403. IEEE (2016)
5. Altieri, M.Á., Nicholls, C.I.: Biodiversidad y manejo de plagas en agroecosistemas, vol. 2. Icaria Editorial (2007)
6. Arboleda, E.R., Fajardo, A.C., Medina, R.P.: An image processing technique for coffee black beans identification. In: 2018 IEEE International Conference on Innovative Research and Development (ICIRD), pp. 1–5. IEEE (2018)
7. Arguello Guadalupe, C.S., León Ruiz, J.E., Díaz Moyota, P.B., Verdugo Bernal, C.M., Caceres Mena, M.E., García Rosero, L.M., Granda de Freitas, J., Saavedra Gallo, C.G.: Incidencia de la cadena de valor en el desarrollo sustentable del cultivo de café robusta (coffea canephora) estudio de caso: Parroquia san jacinto del búa, provincia de santo domingo de los tsáchilas-ecuador. Europ. Sci. J. ESJ **13**(1) (2017)
8. Caicedo-Ortiz, J.G., De-la Hoz-Franco, E., Ortega, R.M., Piñeres-Espitia, G., Combita-Niño, H., Estévez, F., Cama-Pinto, A.: Monitoring system for agronomic variables based in wsn technology on cassava crops. Comput. Electron. Agric. **145**, 275–281 (2018)
9. Ceballos, M.R., Gorricho, J.L., Palma Gamboa, O., Huerta, M.K., Rivas, D., Erazo Rodas, M.: Fuzzy system of irrigation applied to the growth of habanero pepper (capsicum chinense jacq.) under protected conditions in yucatan, mexico. Int. J. Distrib. Sens. Networks **11**(6), 123543 (2015)
10. Culman, M., Portocarrero, J.M., Guerrero, C.D., Bayona, C., Torres, J.L., de Farias, C.M.: Palmnet: an open-source wireless sensor network for oil palm plantations. In: 2017 IEEE 14th International Conference on Networking, Sensing and Control (ICNSC), pp. 783–788. IEEE (2017)
11. Cumbis, E., Jiménez, R.: Análisis Sectorial del Café en la Zona 7 del Ecuador, p. 132 (2012). http://dspace.utpl.edu.ec/bitstream/123456789/2703/1/338X1227.pdf
12. Erazo, M., Rivas, D., Pérez, M., Galarza, O., Bautista, V., Huerta, M., Rojo, J.L.: Design and implementation of a wireless sensor network for rose greenhouses monitoring. In: 2015 6th International Conference on Automation, Robotics and Applications (ICARA), pp. 256–261. IEEE (2015)
13. Erazo-Rodas, M., Sandoval-Moreno, M., Muñoz-Romero, S., Huerta, M., Rivas-Lalaleo, D., Naranjo, C., Rojo-Álvarez, J.: Multiparametric monitoring in equatorian tomato greenhouses (i): Wireless sensor network benchmarking. Sensors **18**(8), 2555 (2018)
14. Fernandez, L., Huerta, M., Sagbay, G., Clotet, R., Soto, A.: Sensing climatic variables in a orchid greenhouse. In: 2017 International Caribbean Conference on Devices, Circuits and Systems (ICCDCS), pp. 101–104. IEEE (2017)
15. Fernandez, L., Huerta, M., Sagbay, G., Clotet, R., Soto, A.: Sensing climatic variables in a orchid greenhouse. In: 2017 International Caribbean Conference on Devices, Circuits and Systems, ICCDCS 2017, pp. 101–104 (2017). https://doi.org/10.1109/ICCDCS.2017.7959719
16. Garcia, D.C., Martínez, G.Á., Garcia, M.E.: Raspberry pi y arduino: semilleros en innovación tecnológica para la agricultura de precisión. Informática y Sistemas: Revista de Tecnologías de la Informática y las Comunicaciones **2**(1), 74–82 (2018)

17. Guambi, L.A.D.: Caracterización física y organoléptica de cafés arábigos en los principales agroecosistemas del Ecuador. INIAP Archivo Historico (2003)
18. Guillermo, J.C., García-Cedeño, A., Rivas-Lalaleo, D., Huerta, M., Clotet, R.: IoT architecture based on Wireless Sensor Network applied to agricultural monitoring: A case of study of cacao crops in Ecuador. In: Second International Conference of ICT for Adapting Agriculture to Climate Change. Springer (2018)
19. Ipanaqué, W., Belupú, I., Castillo, J., Salazar, J.: Internet of things applied to monitoring fermentation process of cocoa at the piura's mountain range. In: 2017 CHILEAN Conference on Electrical, Electronics Engineering, Information and Communication Technologies (CHILECON), pp. 1–5. IEEE (2017)
20. Jiménez, A., Massa, P.: Producción de café y variables climáticas: El caso de Espíndola, Ecuador. Economía **40**, 117–137 (2015). http://www.redalyc.org/pdf/1956/195648804006.pdf
21. Jiménez, M., Jiménez, A., Lozada, P., Jiménez, S., Jiménez, C.: Using a wireless sensors network in the sustainable management of African palm oil solid waste. In: 2013 Tenth International Conference on Information Technology: New Generations (ITNG), pp. 133–137. IEEE (2013)
22. Kodali, R.K., Soratkal, S., Boppana, L.: WSN in coffee cultivation. In: 2016 International Conference on Computing, Communication and Automation (ICCCA), pp. 661–666. IEEE (2016)
23. Laban, N., Abdellatif, B., Ebied, H.M., Shedeed, H.A., Tolba, M.F.: Performance enhancement of satellite image classification using a convolutional neural network. In: International Conference on Advanced Intelligent Systems and Informatics, pp. 673–682. Springer (2017)
24. Lin, B.B.: The role of agroforestry in reducing water loss through soil evaporation and crop transpiration in coffee agroecosystems. Agric. Forest Meteorol. **150**(4), 510–518 (2010). https://doi.org/10.1016/j.agrformet.2009.11.010
25. Mancuso, M., Bustaffa, F.: A wireless sensors network for monitoring environmental variables in a tomato greenhouse **10** (2006)
26. Martins, G.D., Galo, M.L.B.T., Vieira, B.S.: Detecting and mapping root-knot nematode infection in coffee crop using remote sensing measurements. IEEE J. Sel. Top. Appl. Earth Obs. Remote Sens. **10**(12), 5395–5403 (2017)
27. Medina-Meléndez, J.A., Ruiz-Nájera, R.E., Gómez-Castañeda, J.C., Sánchez-Yáñez, J.M., Gómez-Alfaro, G., Pinto-Molina, O.: Estudio del sistema de producción de café (coffea arabica l.) en la región frailesca, chiapas. CienciaUAT **10**(2), 33–43 (2016)
28. Palomino, U., Alexandra, J.: Análisis del impacto de la política económica en el sector agrario del Ecuador 2007 2015 (2017). http://repositorio.utmachala.edu.ec/handle/48000/10509
29. Piamonte, M., Huerta, M., Clotet, R., Padilla, J., Vargas, T., Rivas, D.: WSN prototype for African oil palm bud rot monitoring. In: International Conference of ICT for Adapting Agriculture to Climate Change, pp. 170–181. Springer (2017)
30. Pinto, C., Furukawa, J., Fukai, H., Tamura, S.: Classification of green coffee bean images based on defect types using convolutional neural network (CNN). In: 2017 International Conference on Advanced Informatics, Concepts, Theory, and Applications (ICAICTA), pp. 1–5. IEEE (2017)
31. Ramos, P.J., Avendaño, J., Prieto, F.A.: Measurement of the ripening rate on coffee branches by using 3d images in outdoor environments. Comput. Ind. **99**, 83–95 (2018)

32. Erazo Rodas, M., Sandoval Moreno, M., MuñoZ Romero, S., Huerta, M., Rivas-Lalaleo, D., Rojo-Álvarez, J.: Multiparametric monitoring in equatorian tomato greenhouses (ii): Energy consumption dynamics. Sensors **18**(8), 2556 (2018)
33. Erazo Rodas, M., Sandoval-Moreno, M., Muñoz-Romero, S., Huerta, M., Rivas-Lalaleo, D., Rojo-Álvarez, J.: Multiparametric monitoring in equatorian tomato greenhouses (iii): Environmental measurement dynamics. Sensors **18**(8), 2557 (2018)

Metamorphosis and Dynamics of Inorganic Species in the Waters of the Chone Multiple Purpose Dam, Ecuador

Rosa Arias-Carrera[1], Paola Álvarez-Castillo[1],
Oswaldo Borja-Goyes[1(✉)], Leidy Cajas-Morales[1],
David Carrera-Villacrés[2], Karla Enríquez-Herrera[1],
Tania González-Farinango[1], David González-Riera[1],
Erika Guamán-Pineda[1], Solange Guitiérrez-Puetate[1],
Andrés Moreno-Chauca[1], Paula Montalvo-Alvarado[1],
Richard Rubio-Gallardo[1], Tatiana Sandoval-Plaza[1],
Nelson Unda-León[1], and Paul Velarde-Salazar[1]

[1] Universidad Central del Ecuador, Facultad de Ingeniería en Geología,
Minas, Petróleos y Ambiental (FIGEMPA), Carrera de Ingeniería Ambiental,
Quito, Ecuador
oaborja@uce.edu.ec
[2] Universidad de las Fuerzas Armadas ESPE, Departamento de Ciencias de la
Tierra y la Construcción, Grupo de Investigación en Contaminación Ambiental
(GICA), Sangolquí, Ecuador
dvcarrera@espe.edu.ec
http://www.gica.espe.edu.ec

Abstract. The concentration of chemical species in natural water is determined by numerous physical-chemical phenomena such as mixing, dissolution, suspension, sedimentation, complex reactions, among others, that determine the presence of certain cations and anions in natural water; these processes are altered by changes in temperature, caused by climate change, which modify the mobility and dilution of pollutants and other chemical substances. This work is intended to determine the dynamics and metamorphosis of the concentrations of organic species in water, caused by the presence of Chone Multipurpose Dam and its possible relation with climate change in Chone, through a laboratory experiment causing evaporation and evaluating these changes through the analysis of evolutionary pathways, Piper trilinear diagrams and Gibbs diagrams of water samples taken between 2013 and 2018. According to the evolutionary paths, the water of the PPMCH has a metamorphosis oriented towards the alkaline sulfated route, water typology evolved from bicarbonated calcic and magnesic water to 2018 where chlorine ions predominate. Additionally, in 2018, the concentration of nitrates decreased in the period of low water and rain due to its capacity to become a limiting nutrient. The processes that dominate the surface water chemistry in the rainy season have a predisposition to the mineralization in equilibrium with the rocks; however, the increase of the global temperature can generate a predisposition of waters towards the area.

Keywords: Dam · Water · Metamorphosis · Climate · Ions

© Springer Nature Switzerland AG 2019
J. C. Corrales et al. (Eds.): AACC 2018, AISC 893, pp. 218–233, 2019.
https://doi.org/10.1007/978-3-030-04447-3_15

1 Introduction

1.1 Chemical Species Variation on Superficial Waters

The composition of surface waters depends on the geographical location of the water, that defines the aquatic geochemical processes [1] varying considerably according to the origin of the water and the path it traverses in which it drags sediments and joins other tributaries, in this journey.

Chemical concentration changes constantly by physical and chemical phenomena such as mixture, dissolution, suspension, acid-base, oxidation-reduction and complexation reactions of chemical species [2]. These phenomena are complex and uncontrolled, making chemical characterization of natural water a hard task [3]. Finally, the tributaries are deposited in the final water body, place in which the sedimentation process produced by various factors such as the intervention of colloidal substances carried by river currents cause a variation of inorganic species [4].

1.2 Influence of Climate Change on Water Chemistry

Climate change, which has its origin in global warming, is a reality that is evident in the drastic changes of meteorological phenomena that take place on Earth, some of that weather could affect river flows, modifying the mobility and dilution of contaminants [5]. Additionally nutrient loads could increase due to climate change [6], these changes are mainly related to the air temperature and the increase of extreme hydrological events, temperature is the main factor affecting physical-chemical phenomena and biological reactions in natural waters [7].

1.3 Influence of Dams on Water Chemistry

Dams or reservoirs that retain water, slow down the contribution of mineral waters downstream, this is the reason why the water that comes out from the dam, leaves behind almost all of its sediment load [8], drastically modifying its chemical composition and can also cause scour problems that affect the engineering works of the dams [9]. Although the effects of the upward alterations of a dam are difficult to determine, it is known that the downstream ones have a greater reach (geographically), altering the processes of erosion, deposition of sediments, flooding of plains and water recharge [10].

1.4 Tools for Water Chemical Variation Analysis

For the multicriterial analysis of water metamorphosis, is necessary to consider Piper's trilinear diagram, which is used to represent the proportion of three components in the composition of a substance (in this case the water analyzed). This diagram shows the major cations and anions such as: Ca^{2+}, Mg^{2+}, Na^+, K^+ and SO_4^{2-}, Cl^-, CO_3^{2-} and HCO_3^- respectively [11]. The sum of these components represents 100% of the composition, showing the dominant type of water.

The Gibbs diagram, which uses a graph that allows to determine the concentration of total suspended solids and the ratio of ions (ratio of cations = $Na^+/Na^+ + Ca^{2+}$, ratio of anions = $Cl^-/Cl^- + HCO_3^-$), with the purpose of spatializing a point and locating it in three different zones (predominance of evapotranspiration zone and predominance of precipitation zone and mineralization in balance with the rocks zone) which allows to identify the tendency of increase or decrease of salts concentration in water [12].

For the qualitative determination of the evolutionary pathways of water, an evaporation experiment is carried out on the analyzed waters, according to the models proposed by Hardie and Eugster [13], and by Risacher and Fritz [14].

1.5 Chemical Processes on the Water of Chone Multiple Purpose Dam

The processes that control the chemistry of the surface water of the hydrographic systems that form Rio Grande, where the PPMCH (acronyms of Chone Multiple Purpose Dam) is located, are in the mineralization zone in equilibrium with the rocks, however, in dryness period, the tendency towards the area where evaporation predominates was observed [15]. The inorganic chemical composition of these waters has a tendency to increase over time, due to evaporation predominance [16].

The objectives of this work were to determine the dynamics and metamorphosis of the concentrations of inorganic species in the PPMCH through a laboratory experiment, causing evaporation and checking the diagram of water evolution by evaporation in the PPMCH of water samples taken between 2013 and 2018.

2 Materials and Methods

2.1 Materials

The water sample of the Chone Multipurpose Project (PPMCH) was taken from the Grande river hydrographic micro watershed in order to observe the increase in the inorganic chemical composition of the water. It covers an area of 60 km^2, is located east of the City of Chone, province of Manabí and travels approximately 30 km to join the Mosquito River [17]

The province of Manabí, in the west of Ecuador, has great variability for rainfall that it receives annually. In a period of 20 years the National Institute of Meteorology and Hydrology (INAMHI), in its station located in Chone, it recorded in the year 1998 an approximate value to 2500 mm, whereas, in the year 1996 the value was close to 600 mm, with an annual average close to those of 1200 mm [18].

Simple samples were taken in the multipurpose dam Chone in order to analyze the metamorphosis of the inorganic chemical composition of water in relation with precipitation and evaporation. The parameters measured were pH, electrical conductivity, SDT, sodium cations, potassium, calcium, magnesium; and anions of carbonates, bicarbonate, sulphates, nitrates and chlorides.

With the results obtained were made curves of distribution of the different anions and cations considering the hydrographic characteristics and the different meteorological periods, to observe the behavior of the inorganic concentration of the Waters.

Figure 1 shows the geographic location of the multipurpose Project Chone, which serve as a reference for sampling.

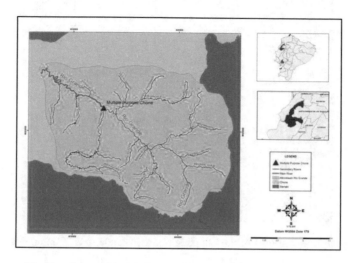

Fig. 1. Geographical location of the sampling point in PPCH.

2.2 Methods

Chone canton, province of Manabí (Fig. 2), has a warm and dry climate from June to November [19]. It has great variability in the amount of rainfall that it receives annually. In a period of 20 years, the National Institute of Meteorology and Hydrology (INAMHI), in its station located in Chone, registered in 1998 an approximate value of 2500 mm, while in 1996 the value was close to 600 mm, with an annual average close to 1200 mm, which hinders the permanent distribution of water demand by the inhabitants and the biota in general [19]. Chone is an area with a high risk of flooding during the rainy season, which is why the PPMC was built, which has been in operation since 2015, located at the confluence of the Rio Grande [20]. According to the Environmental Impact Study carried out for the construction of the Chone Multipurpose project, disseminated by the National Water Secretariat (SENAGUA), in December 2010, data was taken from the M162 Chone station during the second semester of 2004 and it was demonstrated a water deficit of between 200 to 600 mm due to low rainfall (Fig. 3).

Data obtained from the INAMHI station M162 determined the annual and monthly rainfall variations over a period of 20 years, data that was completed and validated in advance with the support of the Portoviejo station.

Water sampling was carried out along the Grande river, based on the INEN 2169: 98 and 2176: 98 standards [19]. The sampling points are located upstream of the PPMCH, and in the Camarones, Cañitas and Cognaque tributaries. 26 points were defined (Fig. 2); the first nine were made in July 2016 and the next 17 points were taken in February 2017 because the hydrographic system of the Grande river showed

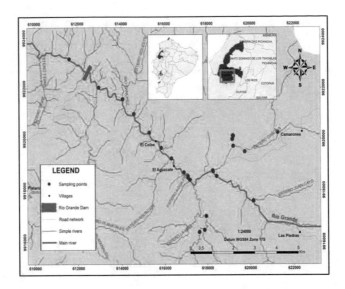

Fig. 2. Location map of the sampling points, Grande river, Canton Chone.

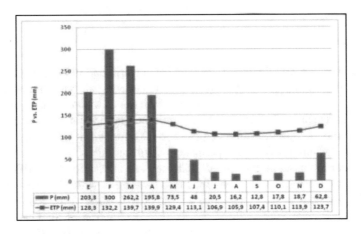

Fig. 3. Climogram of the year 2004 with data from the M162 Chone station. **Source.** SENAGUA, 2010.

an increase in eutrophication [20]. The last point was sampled in January 2018. The physicochemical parameters were analyzed, according to national and international standards (NTE-INEN-0973, EPA 1983, 1701, NOM-AA-93-1984, Standard methods, HACH 2000), in the Environment Laboratory of the University of the Armed Forces - ESPE (Table 1).

The results of the phosphate and temperature levels, pH and EC, were tabulated for drought and rainy season, with the change of the climatic forms, as these influence the concentrations [19, 21]. Piper's trilinear diagram was used to know the composition of the water analyzed based on three components, also the Gibbs diagram to establish the increase or decrease according to the trend of the concentration of salts in the water. Experimentally, the evaporation of the types of water analyzed was used (Table 2).

Table 1. Methods of physical and chemical determinations for water samples.

Parameter	Reference	Method
Temperature (T)	pH meter Hanna Instruments HI 2210	Direct electrode measurement
Potential of Hydrogen (pH)	pH meter Hanna Instruments HI 2210	Direct electrode measurement
Electric Conductivity (CE)	Electrical conductivity meter Thermo Scientific Orion Star A212	Direct electrode measurement
Nitrates (NO_3^-)	Spectrophotometer HACH DR 5000	APHA 8190
Phosphates (PO_4^{3-})	Spectrophotometer HACH DR 5000	APHA 8190
Calcium (Ca^{2+})	Spectrophotometer HACH DR 5000	APHA 3500 Ca B
Magnesium (Mg^{2+})	AA Spectrophotometer PerkinElmer PinAAcle 900T	APHA 3500 Mg B
Sodium (Na^+)	AA Spectrophotometer PerkinElmer PinAAcle 900T	APHA 3500 Na B
Potassium (K^+)	AA Spectrophotometer PerkinElmer PinAAcle 900T	APHA 3500 K B
Carbonates (CO_3^{2-})	Titration	APHA 2320
Bicarbonates (HCO_3^-)	Titration	APHA 2320 B
Chlorides (Cl^-)	Titration	APHA 4500 Cl$^-$ B
Sulfates (SO_4^{2-})	Titration	APHA 8051

Table 2. Locations and coordinates of sampling points

Year	Season	Reference	Coordinates		
			X	Y	Z
2013	Dry	Point in PPMCH	612893	9922166	44
2014	Rain	Point in PPMCH	612893	9922166	44
2016	Rain	Point in PPMCH	612893	9922166	44
2018	Rain	Point in PPMCH	612893	9922166	44

3 Results and Discussion

3.1 Physical-Chemical Analysis

See Tables 3, 4, 5, 6, and 7.

Table 3. Physical-chemical analysis of water

Year	pH	CE (mS/cm)	NO_3^- (mg/L)	TDS ppm	PO_4^{3-} (mg/L)
2013	7.90	0.354	-	324	-
2014	8.10	0.221	-	192	-
2016	6.94	0.234	0.138	226	0.820
2018	7.21	0.224	2.100	301	0.325

3.2 Climatic Data of the PPMCH Area and Its Relationship with Climate Change

Between the years of study, the samples were taken during the dry season (2013 and 2018) and rainy season (2014, 2016 and 2018). Regard to dry season, as shows in Fig. 4, there is a variation of +0.9 °C in 2018 with respect to the monthly average temperature (1985–2009 average) and a variation of −8.9 mm/month of precipitation between 2018 and monthly average precipitation (1985–2009 average). Regard to rainy season, as shows in Fig. 5, there is a variation of +0.2 °C in 2018 with respect to the monthly average temperature (Table 7) and a variation of −143.0 mm/month of precipitation between 2018 and monthly average precipitation (Table 7).

  The tendency of temperature increase of these results is valid since there exists a tendency of increase of temperature of 0.22 °C by decade in the basin of Chone. It is according to a study for the period 1961–2005 realized in the Project Adaptation to Climate Change through Effective Water Governance in Ecuador (PACC), as evidence in the detection of climate change in the region [22, 34]. In addition, according to Contreras and others (2014), evidence was found that climate change is affecting the coastline of the province of Manabí with a change rate of the ambient temperature of 0.012 °C/year. While, the decrease in precipitation according to the figures, is corroborated with the decreasing trends of annual precipitation have been observed on Chile and part of the western coast of South America [23]. Further, it is based in the results obtained by Pourrut and Nouvelot (1995) of the decrease in precipitation in the Ecuadorian coastal region between 1915 and 1983, and with the descent behavior obtained in the Hydrological Study of floods in the Chone basin [24], in the period 1964–2005. Decreasing trends of annual precipitation have been observed on Chile and part of the western coast of South America.

Table 4. Physical-chemical analysis of water: anions and cations

Year	Ca^{2+} (meq/L)	Mg^{2+} (meq/L)	Na^{+} (meq/L)	CO_3^{2-} (meq/L)	HCO_3^{-} (meq/L)	Cl^{-} (meq/L)	SO_4^{2-} (meq/L)
2013	1.47	1.01	0.83	0.00	1.98	0.23	1.175
2014	1.07	0.55	0.54	0.00	1.50	0.11	0.62
2016	0.45	0.24	0.47	0.00	1.84	0.09	0.67
2018	0.113	0.112	0.369	0.00	0.00	0.00	0.00

Table 5. Physical-chemical analysis of the water of evaporated samples

Year	pH	CE (mS/cm)	NO_3^{-} (mg/L)	TDS ppm	PO_4^{3-} (mg/L)
2018	7.10	0.180	1.300	136	0.630
2018	8.40	0.250	1.400	189	0.768
2018	8.50	0.330	1.500	237	0.645
2018	8.50	0.430	1.600	313	0.056

3.3   Concentration Dynamics of Inorganic Species in the PPMCH

Figures 6 and 7 show the change of the different parameters analyzed, between 2013 and 2018. Throughout 2013–2018, except for 2015, both the pH and the electrical conductivity have a balanced behavior, pH does not suffer extreme alterations or changes in concentration for any of the seasons whether these, dry or rain (Table 3). The increase in temperature affects the internal chemical processes of the lakes. There have been decreases in dissolved inorganic nitrogen, an increase in salinity, and pH increases in soft water lakes [23]. There is no trend in the increase of pH in the study waters (Table 3) due to an increase in ambient temperature (Figs. 4 and 5), possibly due to damping factors and chemical equilibria between the water and the lithology of the basin. However, in the experimentation there was an increase in pH as the water evaporated due to an increase in temperature (Table 5). The electrical conductivity presents an increase in the year 2013, corresponding to the dry season, this is related to the temperature and low water level, that

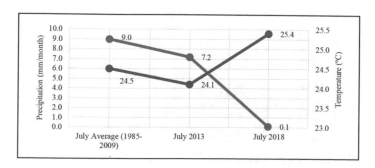

Fig. 4. Comparison of rainfall and temperature data in the dry season (July) between average precipitation, years 2013 and 2018.

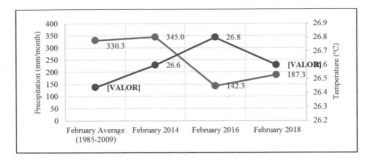

Fig. 5. Comparison of rainfall and temperature data in the rainy season (February) between average precipitation, years 2014, 2016 and 2018.

allowed a greater concentration of components, it is important that EC is related to TDS and temperature, as it relates for this reason for the dry season a decrease in solids is noticeable. While for the rainy season there is a high concentration of TDS, due to that these are the measure of the content of all inorganic and organic substances in a liquid in suspended molecular, ionized or micro-granulated form [25].

Carbonate concentrations are very low (Table 4), this occurs when the pH of the water is acid, and at values higher than 6.5, bicarbonate is the predominant species, as shown in Fig. 6. Although the pH of the water is 6–8.5, the carbonates have almost zero concentrations; this is possibly due to factors such as temperature and the contribution of weak salts such as borates, silicates, nitrates and phosphates in the alkalinity. The bicarbonates (Fig. 6), decrease over the years, with a possible independence of the climatic epoch, perhaps due to multiple factors such as weathering for the contribution of carbonate rocks, the presence of carbon dioxide in water, decomposition of organic matter, among others; that is, water in the reservoir area can have a good buffer capacity [26]. Because of the increase in temperature due to climate change, in some lakes in Europe and East Africa there has been a decrease in nutrients in surface waters, with the consequent increase in concentration in deep waters with a higher

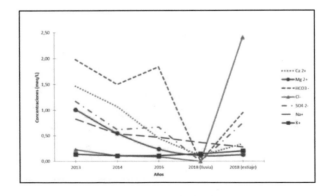

Fig. 6. Variation of ion concentration in PPMCH water between 2013 and 2018.

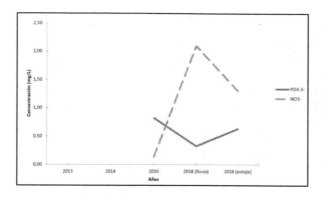

Fig. 7. Variation of phosphates and nitrates in PPMCH water between 2013 and 2018.

concentration of sulphates, basic cations and silica, and a higher alkalinity and conductivity [23]. Thus, the tendency to increase the temperature over the years (Figs. 4 and 5), possibly causes a decrease in sulphates of surface water (Table 4). However, the experimental results (Table 6) show an increase in the concentration of sulfates as the water evaporates due to an increase in temperature. This coincides with the behavior of nutrients in deep waters, where their concentration increases.

Table 6. Physical-chemical analysis of water: anions and cations of the evaporated sample

Year	Ca^{2+} (meq/L)	Mg^{2+} (meq/L)	Na^+ (meq/L)	CO_3^{2-} (meq/L)	HCO_3^- (meq/L)	Cl^- (meq/L)	SO_4^{2-} (meq/L)
2018	0.35	0.13	0.28	0.00	0.95	2.40	0.75
2018	0.35	0.13	0.39	0.40	0.50	3.20	0.98
2018	0.34	0.13	0.49	1.10	0.05	8.60	1.24
2018	0.08	0.95	1.39	0.60	0.45	3.40	1.87

The concentration of chlorides is very low in first years (Fig. 6), however, in the rainy season of 2018, the values increase considerably (from 0.09 to 2.40 meq/L). This is possibly due to natural factors such as washing of halite rocks and silicates, washing of arid lands and even intrusions of seawater; while anthropic sources can be runoff from agricultural and livestock irrigation areas [26].

Variations in the concentration of sulphates have a similar behavior, over the years, to the variations of bicarbonates (Fig. 6), however, there is an apparent relationship with the climatic seasons, since in periods of low water the amounts of sulphates are greater, perhaps because of the processes of evapotranspiration. In addition, these changes are possibly due to the washing of sulphurous minerals as well as the drainage of agricultural irrigation areas [26].

The concentrations of calcium, magnesium and sodium ions are reduced as time passes, and there is no possible relationship with the rainy or dry seasons (Fig. 6). In the case of calcium, this is explained by the fact that it is a non-conservative ion, and possibly the changes appreciated are due to the influence of temperature, pH and

Table 7. Climatic data of the area of the PPMCH.

Year	Month	Season	Precipitation (mm)	Temperature (°C)
1985–2009	July	Dry	9.0	24.5
1985–2009	February	Rainy	330.3	26.4
2013	July	Dry	7.2	24.1
2014	February	Rainy	345.0	26.6
2016	February	Rainy	142.3	26.8
2018	February	Rainy	187.3	26.6
2018	July	Dry	0.1	25.4

concentration of carbon dioxide that affect the solubility [26]. On the other hand, this decrease can be caused by the increase of eutrophication in the reservoir [27], since this increases the photosynthetic rate and water decalcification occurs. The magnesium ion comes mainly from the weathering of rocks and even its increase is due to the intrusion of seawater; therefore, there is possibly a small contribution of local rocks. The decrease in sodium ion is probably due to a reduction in the natural washing of silicates and evaporites, and apparently, there is no anthropogenic contribution from the agricultural sector [26]. The concentration of potassium ions increases slightly in times of low water (Fig. 6), these low concentrations are possibly because the phosphate rocks are resistant to washing and potassium easily recombines with the products of weathering. These changes in sodium, calcium and magnesium ions are narrowed with climate change, because, as with sulfates, as the temperature increases, the cations decrease in the surface water. However, in the experimentation (Table 6), these cations increase, because it coincides with the behavior observed in deep water of lakes and rivers in Europe [23].

3.4   Water Metamorphosis of the PPMCH in an Evaporation Process due to Temperature Increase and Climate Change

In order to predict the behavior and evolution of PPMCH natural waters caused by temperature increase, an experiment was developed with water samples of 2018, measuring the chemical concentration of the ions in the original sample at 1000 ml, and evaporating and measuring the same sample at 750 ml, 500 ml and 250 mL; according to the methods proposed by Hardie and Eugster [13], that stabilised the evolution model of the chemical composition of water by evaporation [28].

Calcite Precipitation
When the water is evaporating, it dissolves the components that are concentrated and precipitate a sequence of minerals increasing solubilities; because of the low solubility the first to be deposited is calcite ($CaCO_3$). During calcite precipitation, the product of the ionic activity remains constant and equal to the product of solubility of calcite. At low salinities, the concentrations vary as the activities, the product of the calcium and carbonate concentrations remain approximately constant. Figure 8 shows that the original water flow of the PPMCH presents a tendency to be in equilibrium

with calcite. However, if it evaporated in ¼, ½ and ¾ of its capacity it would tend to be directed towards the carbonated way. Calcite always has a tendency to precipitate, as in the original water there is more concentration of calcium than carbonate, the solution is concentrated in calcium producing neutral brines with a pH lower than 9 [28] (Fig. 9).

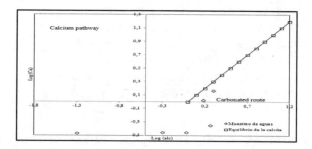

Fig. 8. Relationship between alkalinity, calcium and balance of calcite from water sampling at PPMCH

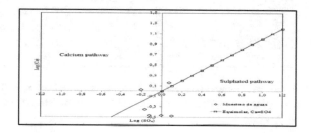

Fig. 9. Evaporitic pathways of the sulphated and calcic evolutionary pathways of the PPMCH

Plaster Precipitation

If at the beginning of the precipitation of the gypsum there is more sulphate than calcium, then the solution will concentrate in sulfate and become impoverished in calcium, obtaining sulfated brines of general type. If there is more calcium than sulfate, the gypsum starts to precipitate and calcium will concentrate, producing calcium brines of the Na-Ca/Cl type. In the end, the obtained results are shown in Table 8.

3.5   Gibbs Diagrams: Change of Water Typology Over Time

Gibbs (1970) identified three natural mechanisms that control the chemistry of the world of superficial continental waters: the mineralization in equilibrium with rocks, predominance and evaporation with regard to precipitation and the predominance of Precipitation with respect to evaporation [29, 31, 32, 33]. The Na: (Na + Ca) relation was calculated to define the behavior that surface water has presented in PPMCH. (Figure 10). The results indicate that in this period the waters have a tendency to the zone of mineralization in equilibrium with rocks. This type of zone is typically located

Table 8. Evolutionary pathways taken by the hydrographic systems of the evaporitic basin of PPMCH

Year	Type of water	Evolution of water by evaporation
2013	Original	Sulfated alkaline or direct route
2014	Original	Sulfated alkaline or direct route
2016	Original	Carbonated route
2018	Original	Calcium pathway
2018	Original	Neutral sulphated pathway
2018	Evaporated 250	Carbonated route
2018	Evaporated 500	Neutral sulphated pathway
2018	Evaporated 750	Direct alkaline sulfated route

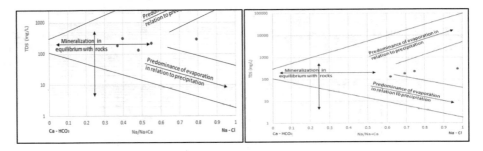

Fig. 10. Gibbs Diagram with annual data for 2013, 2014, 2016 and 2018; and Gibbs diagram with 2018 data and samples evaporated at 750, 500, 250 mL (from left to right).

in the Tropic of South America, with characteristics of poor soils by the washing of minerals [16]. It is important to emphasize that in the sample taken in June 2018 is tended to the predominance of precipitation with respect to evaporation, which shows greater concordance because in the equatorial area where this area is located rainfall is high in Comparison with the rest of the planet [30].

3.6   Piper's Diagrams: Change in Water Typology Over Time

The obtained result in the first diagram of Fig. 11 evidences a metamorphosis of the water due to the PPMCH, where the water in 2013, 2014 and 2016 was characterized as a bicarbonated calcic and magnesic. In 2018, there is a predominance of chloride ions, causing the water to be classified as chlorinated and sulphated sodium. With the simulation of an abrupt reduction of the volume of water due to evaporation caused by a possible climate change, the second Piper diagram of Fig. 11 is obtained. Because of the evaporation of water, certain ions predominate, in this case the chlorine ion, this accentuates the characterization of a sodium sulphated chloride water.

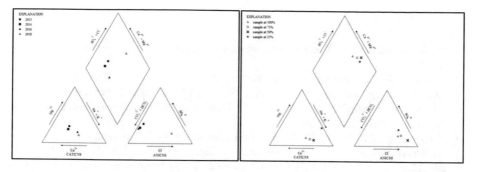

Fig. 11. Piper's diagram of the water samples taken in 2013, 2014, 2016 and 2018; and Piper's diagram of the water samples taken in 2018 (from left to right).

4 Conclusions

According to the evolutionary paths, PPMCH water has a metamorphosis oriented towards the alkaline sulfated route in which the solution begins its evolution through the carbonated route, but the precipitation of magnesium silicates cuts this path, allowing calcium to concentrate until reaching the precipitation of the plaster. At this point, sulfate predominates over calcium and a sulphated brine is obtained.

The processes that dominate the surface water chemistry in PPMCH in the rainy season, have a predisposition to the mineralization in equilibrium with rocks, however, the increase of the global temperature can generate a predisposition of waters towards the area, where evaporation predominates with respect to precipitation, which causes a rise of the inorganic chemical composition of the waters.

The results show a water metamorphosis over the years due to PPMCH, in addition to simulate a future scenario of river water evaporation due to climate change. The bicarbonated calcic and magnesic water (2013) changed into a water where chlorine ions predominate (2018). Additionally, the experiment shows that if this process continues, there will be even greater amounts of chlorine in relation to other ions present in water.

In 2018, the concentration of nitrates decreased in the period of low water and rain due to its capacity to become a limiting nutrient, and for this reason the influence of other components affects and reduces its range. The concentration of phosphates also decreases due to the low alkalinity content of the PPMCH area. The concentrations of carbonates and bicarbonates are possibly affected by the dissolution of rocks, and the water could have a buffering capacity. The variations of calcium, magnesium, sodium and potassium ions are explained by the possible natural causes as a contribution of rock washing and even marine intrusion, as well as the increase in eutrophication that affects the calcium ion, the concentrations of chloride and sulfate ions are affected by natural and anthropic causes, even during the climatic periods.

References

1. Vijay, S.: Handbook of Applied Hydrology, 2nd edn. McGraw Hill Professional, New York (2016)
2. Manahan, S.: Environmental Chemistry, 10th edn. CRC Press, Boca Ratón (2017)
3. Benjamin, M.: Water Chemistry, 2nd edn. Waveland Press, Long Grove (2014)
4. Álvarez, J., Asedegbega, E., Muñoz, V.: Gestión y conservación de aguas y suelos. Editorial UNED, Madrid (2017)
5. Battarbee, R., Kernan, M., Wade, A., Whitehead, P., Wilby, R.: A review of the potential impacts of climate change on surface water quality. Hydrol. Sci. J. **54**(1), 101–123 (2009)
6. Bidoglio, G., Bouraoui, F., Galbiati, L.: Climate change impacts on nutrient loads in the Yorkshire Ouse catchment (UK). Hydrol. Earth Syst. Sci. Dis. **6**(2), 197–209 (2002)
7. Baures, E., Clement, O., Delpla, I., Jung, A.: Impacts of climate change on surface water quality in relation to drinking water production. Environ. Int. **35**, 1225–1233 (2009)
8. García, J., Puigdefábregas, J.: Efectos de la construcción de pequeñas presas en cauces anastomosados del pirineo central. Cuadernos de Investigación Geográfica, vol. 11, no. 1–2, pp. 91–102 (1985)
9. Park, C.: Man-induced changes in stream channel capacity. In: River Channel Changes, pp. 121–141 (1977)
10. Calmus, T., Búrquez, A., Yrízar, A.: Disyuntivas: impactos ambientales asociados a la construcción de presas. Región y Sociedad, vol. 24, no. 3, pp. 289–307 (2012)
11. Piper, A.: A graphic procedure in the geochemical interpretation of water-analyses. Trans. Am. Geophy. Union **25**(6), 914–928 (1944)
12. Gibbs, R.: Mechanisms controlling world water chemistry. Science New Ser. **170**(3962), 1088–1090 (1970)
13. Hardie, L., Eugster, H.: The evolution of closed basin brines. Mineral. Soc. Am. Spec. Paper **3**, 273–290 (1970)
14. Risacher, F., Fritz, B.: Origin of salts and brine evolution of Bolivian and Chilean Salars. Aquat. Geochem. **15**, 123–157 (2009)
15. Carrera-Villacrés, D., Guevara-García, P., Gualichicomin-Juiña, G., Maya-Carrillo, A.: Processes controlling water chemistry and eutrophication in the basin of Río Grande, Chone, Ecuador. In: Ciencias Químicas y Matemáticas Handbook T-I, pp. 25–36 (2015)
16. Carrera-Villacrés, D., Crisanto, T., Guevara, P., Maya, M.: Relationship between the in organic chemical composition of water, precipitation and evaporation in the basin of Rio Grande, Chone, Ecuador. Enfoque **6**(1), 25–34 (2015)
17. Anchundia, A.: Población Río Grande frente al proyecto de Propósito Múltiple Chone (2012)
18. Carrera-Villacrés, D., Crisanto-Perrazo, T., Guevara, P., Maya, M.: Relación entre la composición química inorgánica del agua, la precipitación y la evaporación en la cuenca de Río Grande, Chone, Ecuador, Quito, Ecuador (2015)
19. Calvopiña, K.: Design of technosols for the retention of phosphates in water, of the Chone Multiple Purpose dam (PPMCH), from soils samples of the Chone canton, Earth and Constructions Sciences, ESPE (2017)
20. Carrera-Villacrés, D., Guevara, P., Gualichicomin, G.: Quality of the waters of the Multiple Chone Purpose Dam, Quito, Pichincha (2015)
21. Bricker, O., Jones, B.: Main factors affecting the composition of natural waters. In: Trace Elements in Natural Waters. CRC Press, Boca Raton (1995)

22. MAE: Estudio de vulnerabilidad actual a los riesgos climáticos en el sector de los recursos hídricos en las cuencas de los Ríos Paute, Jubones, Catamayo, Chone, Portoviejo y Babahoyo. Manthra Editores, Quito (2009)
23. IPCC: El Cambio Climático y el agua, Ginebra (2008)
24. INAMHI y FAO: Estudio Hidrológico de inundaciones en la Cuenca Alta del Río Chone, Quito (2008)
25. Valles-Aragón, M.C.: Calidad del agua para riego en una zona nogalera del estado de chihuahua. In: SciELO, p. 15 (2017)
26. Betancourt, C., Labaut, Y.: La calidad físicoquímica del agua en embalses, principales variables a considerar. Revista Agroecosistemas, vol. 1, n° 1, pp. 78–103 (2013)
27. Carrera, D., Guerrón, E., Cajas, L., González, T., Guamán, E., Velarde, P.: Relación de temperatura, pH y CE en la variación de concentración de fosfatos, Revista Congreso de Ciencia y Tecnología, vol. 13, pp. 37–40 (2018)
28. Carrera, D.: Salinidad en suelos y aguas superficiales y subterraneas de la cuenca evaporitica de Río Verde-Matehuala, San Luis Potosí. Investigación en Ciencias Agricolas, pp. 122–139 (2011)
29. Gibbs, R.: Mechanisms controlling world water chemistry. Science **170**, 1088–1090 (1970)
30. Aparicio, F.: Fundamentos de hidrología de superficie. LIMUSA, México (2012)
31. Risacher, F., Alonso, H., Salazar, C.: Geoquimica de aguas en cuencas cerradas I, II y III regiones Chile, Chile (1998)
32. Contreras López, M., Cevallos Zambrano, J., Erazo Cedeño, T., Alday González, M., Mizobe, C.: Cambio y variabilidad climática contemporáneos en la costa de Manabí, Ecuador, La Técnica, pp. 90–99 (2014)
33. Pourrut, P., Nouvelot, J.-F.: Anomalías y fenómenos climáticos extremos, de El agua en el Ecuador. Clima, precipitaciones, escorrentía, Quito, Corporación Editora Nacional, pp. 67–76 (1995)
34. Carrera, D., Crisanto, T., Guevara, P., Maya, M., Relationship between the inorganic chemical composition of water, precipitation and evaporation in the basin of Rio Grande, Chone, Ecuador, Enfoque UTE, V.6-N.1, pp. 25–34 (2015)

Evaluation of Threats to Agriculture in the Totaré River Basin Due to Changes in Rainfall Patterns Under Climate Change Scenarios

Freddy Duarte$^{(\boxtimes)}$ ⓘ, Jordi Rafael Palacios ⓘ,
and Germán Ricardo Santos ⓘ

Escuela Colombiana de Ingeniería Julio Garavito, Bogotá, Colombia
`freddy.duarte@escuelaing.edu.co`

Abstract. In this paper we analyze the implications of different representative concentration pathway scenarios (RCP2.6, RCP4.5 and RCP8.5) on the precipitation of the Totaré River basin located in the Department of Tolima, Colombia, and its possible consequences on the productive systems of the area. In the analysis, we employed the global climatic model MPI-ESM-MR with 22 pluviographic stations of the IDEAM and the CSD (Chaotic Statistical Downscaling), a novel downscaling model for different intervals of accumulation of precipitation (5, 7, 10, 15 and 30 days). The results predict an increase of the precipitation in the Totaré River Basin from 10% to 50% for the middle and end of the century under all RCP scenarios. It is necessary to take measures to ensure adequate agricultural production due to possible flooding or soil erosion.

Keywords: Agriculture · Precipitation · GCM

1 Introduction

Climate change, due to natural or anthropogenic causes, can generate serious consequences for the environment and humanity. Therefore, in order to prevent and mitigate the negative impacts of this phenomenon in the environment, including its effects on the hydrological cycle, on a particular region, it is of great importance to improve predictions founded on state of the art techniques.

We evaluated the socio-economic impacts of climate change in the region considering future scenarios of greenhouse gas emissions established by the IPCC (Intergovernmental Panel on Climate Change) with climate projections made with GCMs (General Circulation Models). Figure 1 shows a schematic of the climate projection process.

However, global climate models present information at spatial and temporal scales greater than those required by hydrological models and are therefore inadequate to assess the local effect of climate change. It is then necessary to carry out a process of change of scale on the projections of the local climate; this process is known as downscaling.

© Springer Nature Switzerland AG 2019
J. C. Corrales et al. (Eds.): AACC 2018, AISC 893, pp. 234–248, 2019.
https://doi.org/10.1007/978-3-030-04447-3_16

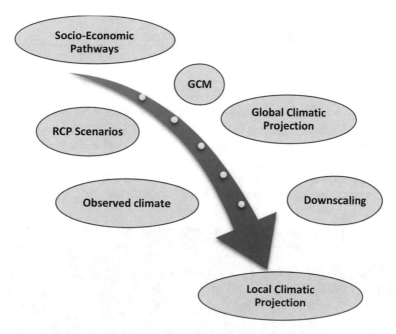

Fig. 1. Climate projection process

Different techniques to transfer the information of fundamental variables from the GCM information to a smaller basin scale have been proposed. These techniques, range from physically based models to stochastic models. Among the most used techniques are: ANN (Artificial Neural Network) [12], SDSM (Statistical Downscaling Model) [13], ADC (Advanced Delta Change Method) [7] and WRF (The Weather Research and Forecast model) [10].

Nevertheless, considering that the climate systems and its associated dynamical processes are essentially non-linear, and possibly chaotic, the effectiveness of these techniques may be limited, and the deterministic estimation of the precipitation obtained from GCMs for local modeling is difficult, as described in [11]. Based on this, the CSD downscaling technique determines the possible presence of deterministic chaos in the dynamical systems, to consider the dynamical behavior in the statstical analysis [3].

Current events have led to an optimal scenario for the creation of different tools to estimate the possible effects of climate change on agricultural production systems. Firstly, due to the global concerns of the increase in frequency and intensity of extreme events, there has been an increase in efforts to development accurate global climate models. Additionally, the management and planning of water resources in different parts of the world has reached a stage where the analysis of possible future changes in the climatology of the region is required. Thirdly, Different international organizations are periodically publishing freely the information results of possible scenarios of climate change and their effects on the production and yield of crops. One such case is presented in [5] where a diagnosis of the main agroclimate models used in Latin America was made, and their benefits and limitations analyzed.

2   Case Study

The Totaré River originates in the central part of Colombia in South America, in the department of Cundinamarca. It is basin covers 6.1% of Tolima's territory with an area of 1428 km^2. Figure 2 shows the general location of Totaré River Basin. The hydrological regime of the river in its main channel is bimodal with an average flow of 19.437 m^3/s, with wet seasons between March-May and September-November, and dry seasons between December-February and June-August.

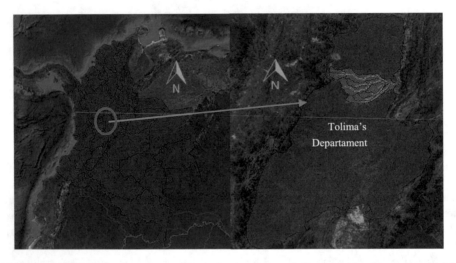

Fig. 2. Location of the Totaré River Basin in the country (left) and in the department (right)

The MPI-ESM-MR model of the Max Planck Institute was used as the Global Climate Model in its first historical realization (1850–2005) with the scenarios RCP2.6, RCP4.5 and RCP8.5 (2006–2100) for a total of 91676 daily precipitation data per scenario, based on the report of the Magdalena River Basin [1]. It was concluded that this GCM reflects adequately most of the local climate state patterns for the study region.

In the present study one cell of the global model was used to represent the study area (as it is presented in Fig. 3), with an average value of 5.13 mm, an average standard deviation of 6.12 mm and an average maximum value of 45.82 mm for the RCP 8.5.

The observed data for the model calibration and validation, was obtained from 22 ground rainfall gauges stations selected from IDEAM (Institute of Hydrology, Meteorology and Environmental Studies, Colombia). The location of the stations is shown in Fig. 4. Data from 1958 to 2016 was processed. The average number of daily precipitation data per station was 13884. The average value was found to be of 4.54 mm, with an average standard deviation of 10.15 mm and an average maximum value of 149.9 mm.

Fig. 3. Cells of the Global Climate Model: MPI-ESM-MR, in the Totaré River basin. Coordinate reference system: WGS-84

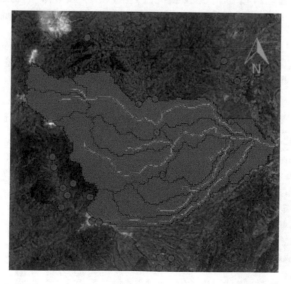

Fig. 4. Ground rainfall gauge stations selected for the downscaling in the Totaré River Basin

The existing systems of agricultural production in the Totaré River basin, as presented in Table 1, was obtained from the Management and Land Use Plan for the Totaré River Basin [2].

Table 1. Predominant production systems identified in the Totaré River basin taken from [2]

Sector of the basin	Production systems
Upper basin	Cold climate production system in hillside soils. Small and medium producers of potato crops and dual-purpose livestock Cold climate production system in hillside soils. Small and medium producers of pea and dairy and meat cattle Average climate production system in hillside soils. Small and medium producers of panela cane
Middle basin	Average climate production system in hillside soils. Small and medium producers' coffee, plantain, corn, bean, avocado and double purpose cattle raising Average climate production system in hillside soils. Small and medium producers of panela cane Warm climate production system in flat and undulating soils. Small and medium producers of rice and dual-purpose livestock
Lower basin	Average climate production system in hillside soils. Small and medium producers of coffee, plantain, corn, bean, avocado and double purpose cattle raising

The main land uses in the Totaré River basin correspond to pastures and forests, as is presented in Table 2, which correspond to 66% of the basin area, while the productive crop area is 21423 Ha, corresponding to 15% of the basin. The productive crop area is farther divided into two types: Semi-permanent or permanent crops corresponding to coffee, plantain, panela and fruit trees, and semi-annual or annual crops represented by oats, peas, rice, vegetables, potatoes and sorghum.

Table 2. Land use covertures in the Totaré River basin taken from [2]

Land use	Area (Ha)	Area (%)
Pastures	70.832,62	49,58
Forest	24.643,57	17,25
Shrub vegetation	17.917,96	12,54
Semi-permanent or permanent crops	11.675,62	8,18
Semiannual or annual crops	9.748,25	6,82
Areas without agricultural and forestry use	6.784,2	4,74
Areas without information	1.270,62	0,89
Total	142.872,84	100

According to [2] rice and coffee are the main water consumers, as shown in Table 3, while water consumption by other crops as cotton, oal, panela, potatoes and peanut, are negligible in comparison.

Table 3. Main water demands in the agricultural sector of the Totaré River basin taken from [2]

Crop	Million m^3/year
Coffee	279.05
Rice	1057,80
Vegetables	0,34
Fruit trees	0,23

Likewise, the following characteristics were also found in the basin management plan:

- A yield of 0.14 L/s-Ha, which corresponds to a low water yield.
- A water scarcity index of 2.24, which indicates a very high demand of water with respect to supply, corresponding to the over-demand in the medium-low basin, in which irrigation areas are included.
- An aridity index from 38 to 48% in the lower basin, exhibiting a large water deficiency. On the other hand, the aridity index is zero in the middle and upper basins, which represents no deficiency of water.
- Temperature-humidity index in the lower basin from 0 to 1.8%, and below 50% in the middle and upper basin.

3 Methods

The Chaotic Statistical Downscaling model (CSD) presented in [3] and evaluated in [4], is a new statistical downscaling method based on the phase space reconstruction for different time steps, identification of deterministic chaos and a general synchronization predictive model. A general scheme of the downscaling method process and related functions is presented in Fig. 5.

We reconstructed the phase space of the dynamic system for both time series (rainfall gauge stations and the global climate model) by the Time-Delay method. It finds the appropriate values of the time delay (τ) and embedding dimension (m) to capture the evolution of the system in a region of attraction. For the selection of the time delay we used the autocorrelation function and the mutual information. was selected using We implemented the correlation dimension and the False Nearest Neighbors (FNN) methods to select the embedding dimension based on the success of previous works [9].

Non-linear Identification	Phase space reconstruction	Type of Motion	General Synchronization

•Power Spectrum
•Kolmogorov Entropy
•Hurst Exponent
	•Time-Delay:
	  •ACF
	  •Mutual information
	•Embedding Dim:
	  •False nearest neighbors
		•Lyapunov exponents spectrum using the algorithm of Eckmann	•Mutual False nearest Neighbors

Fig. 5. Conceptual scheme of the functions and steps in the CSD model

We assessed the presence of deterministic chaos and the identification of the type of movement of the dynamic system for different accumulation intervals by calculating the Lyapunov exponent for both the GCM and the local-observed systems. Then we constructed a chaotic predictive model with the time delay, the embedding dimension and the Lyapunov exponents found for the "optimal and predictable" precipitation accumulation interval of each dynamic system.

In the last decade, the synchronization of dynamical systems has been used successfully in practical and experimental dynamical non-linear systems such as electrical and electronic circuits, lasers, telecommunications systems and chemical reactions [6]. The synchronization of two or more dynamical systems is a fundamental phenomenon that occurs when at least one of the systems changes its trajectory due to a unidirectional or bidirectional coupling with another system, allowing a coherent behavior of the coupled systems.

The existence of a function that relates the variables of two systems suggests that this function can be found and used for prediction purposes. In effect, if two systems are synchronized, then the state of the "response system" can be predicted only by the state of the "driver system", even if the relationship between the variables is unknown.

The predictive function of the CSD model is based on the general synchronization between two dynamical systems: the drive one is the GCM and response is the information of the rainfall gauge stations), characterized by the parameter μ of the mutual false nearest neighbor's method (MFNN). The mutual false nearest neighbor's method created by [8] is a statistical technique based on the calculation of the parameter μ, which evaluates the local neighborhoods between two-time series. Figure 6 shows the relation between points of the two attractors for this method

The CSD model has been used with relative success to predict both precipitation and temperature in different basins of Colombia, such as the Bogotá river basin and the Guali river basin. As presented in [3, 4], in these basins the method has also been compared with other downscaling techniques such as: k-NN Bootstrapping, Delta Change Method, Analogous Methods, Weather Generators and Generalized Linear Methods under three different measures of error, among them the Root-Mean Square Error (RMSE).

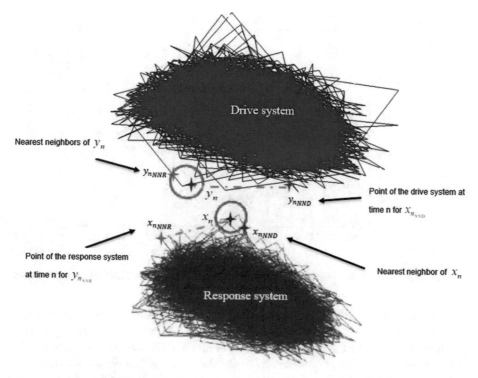

Fig. 6. Relation between nearest neighbors of drive and response systems for general synchronization with mutual false nearest neighbors

For the evaluation of the presence of deterministic chaos in the time series of the Totaré River basin, the following accumulation intervals of precipitation were used: 5 days, 7 days, 10 days, 15 days and 30 days. Likewise, the following dates of calibration, validation and application of the scale reduction technique were used:

1. Calibration: February 1, 1958 to December 31, 2005.
2. Validation: January 1, 2006 to August 30, 2016.
3. Application: August 1, 2016 to December 31, 2100.

4   Results

We found that in both, the GCM's cells and the rainfall gauge stations time series, starting from a rainfall accumulation of 5 days the type of movement of the dynamic system is no longer random (noise) and becomes mainly deterministic chaos, ensuring the short-term predictability for climate projections. Additionally, it was possible to compare the annual precipitation of the GCM with the rainfall gauge stations. A process of bias correction was necessary due to the significant difference between the average values of the two series, as it is presented in Fig. 7.

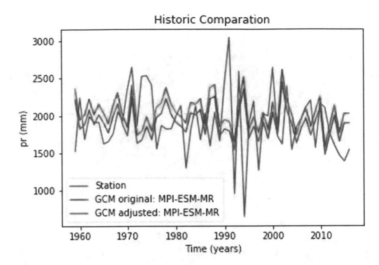

Fig. 7. Comparison of the annual precipitation for the GCM and the 21210030-rainfall station.

As part of the analysis, the monthly multiannual precipitation was compared for a wide variety of scenarios (RCP2.6, RCP4.5 and RCP 8.5) with the reference period, as presented in Fig. 8 for the station 21210020. In all the RCP scenarios the precipitation in the Basin is the highest in the months of October and April, while the month of July is the driest. It was also found that the monthly multiannual precipitation curve will increase in all scenarios.

It was possible to compare the future annual precipitation projections of all the future scenarios. The results of the projections for station 21210020 are shown in Fig. 9. It was found that the RCP 4.5 scenario presents the greatest variability with respect to the mid-century mean value. Nevertheless, a similar pattern was observed in all projected cases; the average annual value of the precipitation increases until it reaches a constant value around 2025.

The process for the validation of the model consisted on graphing the precipitation projections of the CSD model and the observed data for the validation, as well as the calculation of the RMSE error. The results of this procedure for station 21240080 are shown in Fig. 10. The RMSE values calculated for this station were 93.55 for scenario RCP 2.6, 94.31 for RCP4.5 and 95.31 for RCP8.5.

Figure 11 shows that under all the different representative concentration scenarios considered in this study, RCP 2.6, 4.5 and 8.5, there is an increase in precipitation. On average it is estimated that there will be a 30% increase in the precipitation in the basin at the middle and end of the century. Although that figures rises up to 50% in scenario 8.5 at the end of the century in the north central part of the basin.

If an increase in precipitations occurs as predicted by all RCP scenarios considered, the productive systems of the Totaré River basin will be affected. It's observed that, for the scenarios analyzed, there will be an increase in precipitation of 25 to 35% at the middle and end of the century in the upper basin with respect to the historical period. A consequence of this increase (in function of the future evapotranspiration) would be

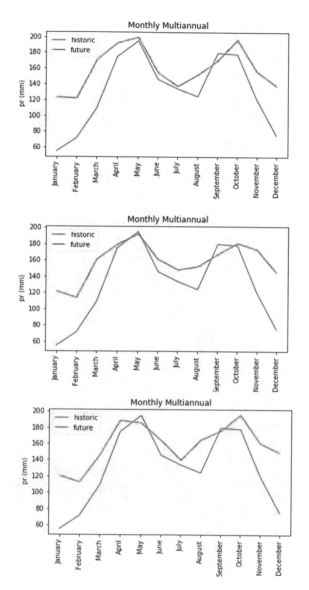

Fig. 8. Comparison of the monthly multiannual precipitation in the Totaré River basin, with respect to the reference period for the three established scenarios RCP2.6 (First row), RCP4.5 (Second row) and RCP8.5 (Third row) in the 21210020-rainfall station.

the associated increase in the humidity index (which currently exceeds 50%). This will, in turn, improve the crop yields and decrease the scarcity index.

Similarly, an increase in precipitation in the middle basin of 20 to 50% at the middle of the century and from 20 to 45% at the end of the century is predicted with respect to the historical period. The consequence of this change in precipitation would

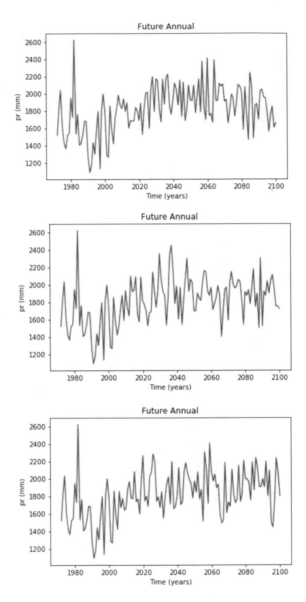

Fig. 9. Comparison of the future annual precipitation in the Totaré River basin, for the three established scenarios RCP2.6 (First row), RCP4.5 (Second row) and RCP8.5 (Third row) in the 21210020-rainfall station.

mirror the situation expected in the upper basin; an increase in the humidity index, which results in the improvement in crop yields, especially in coffee crops (given its high-water demand), and a decrease in the current scarcity index. It is important to highlight that the greatest variations among the results of evaluating the different RCP scenarios occurs in this area.

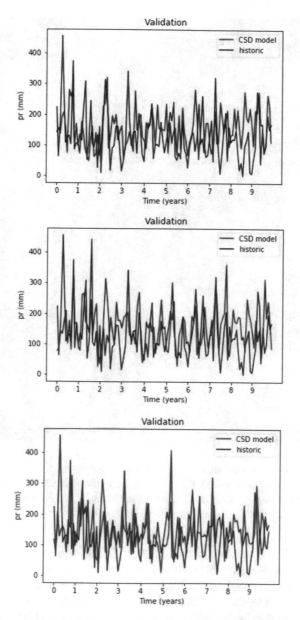

Fig. 10. Validation of the CSD Model for precipitation in station 21240080, in the scenarios: RCP2.6 (First row), RCP4.5 (Second row) and RCP8.5 (Third row).

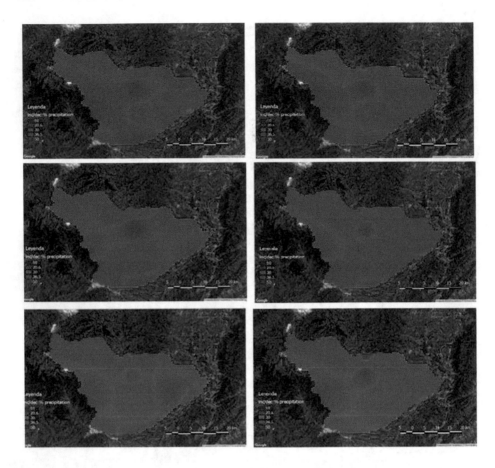

Fig. 11. Comparison of the increase in precipitation in the Totaré River basin, with respect to the reference period for half a century (2035–2065) (left column) and end of the century (2070–2100) (right column) for the three established scenarios RCP2.6 (First row), RCP4.5 (Second row) and RCP8.5 (Third row). Reference values: Blue: 50%, Yellow: 30%, Red: 10%.

The estimated increase in precipitation in the lower basin ranges from 35 to 40% at the middle of the century, and from 20 to 40% at the end of the century with respect to the reference period. The effects of this change would be expected to parallel those in the upper and middle basin, with the additional consideration that there are fruit trees plantations in the lower basin, whose high-water demand could be satisfied.

In general, the expected increase in precipitation along the basin in future years could satisfy the high water demands of the rice and coffee crops, two crops of great economic importance for the region, as well as decreasing the scarcity and aridity indexes.

It is also important to note that according to [2]: "22.51% of the basin is located by pastures that have appeared after leaving the ground without forest cover which makes the soil behave like a waterproof surface making it difficult to infiltration, increasing erosion and decreasing the availability of water for the inhabitants of the whole Basin in general". The characteristics mentioned beforehand make the basin especially susceptible to the predicted increase in precipitation. There could be risks of floods and the erosions of natural soils.

5 Conclusions

Based on the projections obtained for the different scenarios studied, a significant increase in rainfall is expected for the end and middle of this century in the Totaré River basin. It is, therefore, very important to carry out basin planning that explicitly considers these changes, in order to carry out an optimal decision making in the agricultural sector. In addition, it is concluded that even though the values of precipitation will increase, the bimodal regime of monthly precipitation under different climatic scenarios will not be modified.

It's concluded that despite the differences in the RCP scenarios, similar spatial patterns are observed in the future in regards to average precipitation in the Totaré river basin. The average increase in precipitation in the middle and end of the century is estimated at 30%, with a difference between the RCP scenarios of 2%. However, the difference in the projected extreme values of precipitation (minimum and maximum) for the different scenarios varies between 10 to 36%.

With respect to the results of the validation process of the CSD downscaling technique, it was found that the method represents adequately the average values of precipitation but fails to accurately predict some of the extreme values in the time series.

It was also found that the most critical scenario corresponds to RCP 8.5, the CSD model predicted in some sectors of the upper-middle part of the Totaré River basin a 50% increase in precipitation. If that scenario were to be true, it would possibly be necessary to implement certain structural measures to protect the crops and productive land from flooding events. Nevertheless, the high-water requirements of different crops in the basin, among them coffee and rice, and the high scarcity indexes, this extreme scenario would be the most favorable in future years for the productive sector, especially in the lower part of the basin.

Finally, it is concluded that the use of GCMs outputs including scenarios based on the Representative Concentration Pathways (RCP), and the use of downscaling techniques (ej. CSD), effectively allow to represent future climate change scenarios at the regional scale. Based on the regional models, it is further possible to quantify the impacts of the changing climate in the productivity of the agricultural sector and allow the planning of mitigation and adaptation measures to minimize the negative impact of this changes.

References

1. Angarita, H.: Methodology to include climatic variability and climate change scenarios in the WEAP model of the Magdalena River Basin macro and results of the simulations. Informe TNC, Bogotá (2014)
2. Corporación Autonoma Regional del Tolima: Management Plan for the Totaré River Basin. CORTOLIMA (2012)
3. Duarte, F., Corzo, G., Hernández, O., Santos, G.: Chaotic Statistical Downscaling (CSD): application and comparison in the Bogotá River Basin. In: La Loggia, G., Freni, G., Puleo, V., De Marchis, M. (eds.) HIC 2018. EPiC Series in Engineering, vol. 3, pp. 626–634. Palermo (2018)
4. Duarte, F., Corzo, G., Hernández, O., Santos, G.: Evaluación Hidrológica de la Tecnica de Reducción de Escala CSD (Chaotic Statistical Downscaling) en la Cuenca del Rio Bogotá. In: XXVIII Congreso Latinoamericano de Hidráulica e Hidrologia, Buenos Aires (2018)
5. Fernandez, M.: Agroclimatic risk assessment by sector. IDEAM (2013)
6. González-Miranda, J.: Synchronization and Control of Chaos. Imperial College Press, Barcelona (2004)
7. Kraaijenbrink, P.: Advanced Delta Change Method Extension of an application to CMIP5 GCMs. Internal report IR 2013–04, Royal Netherlands Meteorological Insitute, De Bilt (2013)
8. Rulkov, N.F., Sushchik, M.M., Tsmiring, L.S., Abarbanel, H.D.I.: Generalized synchronization of chaos in directionally coupled chaotic systems. Phys. Rev. E **51**(2), 980–994 (1995)
9. Siek, M.: Predicting Storm Surges: Chaos, Computational Intelligence, Data Assimilation, Ensembles. Delft (2011)
10. Soares, P., et al.: WRF high resolution dynamical downscaling of ERA-Interim for Portugal. Clim. Dyn. **39**(9–10), 2497–2522 (2012)
11. Sivakumar, B., Berndtsson, R.: Advances in Data-Based Approaches for Hydrologic Modeling and Forecasting. World Scientific Publishing, Singapore (2010)
12. Vu, M., et al.: Statistical downscaling rainfall using artificial neuronal network: significantly wetter Bangkok? Theoret. Appl. Climatol. **126**(3–4), 453–467 (2016)
13. Wilby, R., Dawson, C.: The Statistical DownScaling Model: insights from one decade of application. Int. J. Climatol. **33**, 1707–1719 (2013)

IoT Network Applied to Agriculture: Monitoring Stations for Irrigation Management in Soils Cultivated with Sugarcane

Edgar Hincapié Gómez, Juliana Sánchez Benítez[✉],
and Javier Alí Carbonell González

CENICAÑA Colombian Sugarcane Research Center, Cali, Colombia
{ehincapie, jsanchez, jacarbonell}@cenicana.org

Abstract. One of the essential practices in sugarcane crops in the Cauca river valley is the water application through irrigation. As the water supply from rainfall is not enough or its distribution is not appropriate to satisfy the water requirements of the crop and since the availability of water is reduced during the two seasons of low rainfall in the Cauca river valley, it is necessary to develop strategies to improve irrigation management. Timely programming of irrigation using sensors to monitor the matric potential of the soil is a strategy that contributes to this objective. In the Research Center of Sugar Cane of Colombia (Cenicaña) research has been undertaken to allow the development of technologies for the use of sensors for the irrigation management in sugarcane crops. The technology is composed by sensors inserted in the soil, the acquisition system, the storage and transmission of data, and the data management and visualization application. In addition, the criteria that must be taken into account for the use of this technology were determined. In a pilot phase, an IoT network was installed in fields cultivated with cane in the Cauca river valley, in order to measure the matric potential of the soil and use it for irrigation scheduling. The results obtained allowed to establish that the threshold to begin the cane crops irrigations is of -85 kPa. It was determined that the matric potential sensors are useful devices for the cane irrigation control, which can be permanently installed in the field to take information manually or automated.

Keywords: Matric potencial of the soil · Sugarcane · Internet of the Things

1 Introduction

In general terms, maintaining an adequate range of moisture content in the soil is a determining factor for agriculture. This is closely related to essential biophysical processes that occur in the soil such as seed germination, plant growth and nutrition, microbial activity and mineralization of organic matter, availability of nutrients in the root zone, and its optimal assimilation. In addition, it is related to the transfer of heat, water and oxygen at the soil-atmosphere interface [1]. Therefore, establishing the availability and quantity of water present in the soil matrix and maintaining it in an

© Springer Nature Switzerland AG 2019
J. C. Corrales et al. (Eds.): AACC 2018, AISC 893, pp. 249–259, 2019.
https://doi.org/10.1007/978-3-030-04447-3_17

optimum range contributes to the productivity and profitability of crops and to water conservation. The determination of water availability in the soil is carried out by the use of matric potential sensors while the amount of water contained in a soil is established by the use of humidity sensors [2].

The water extraction that the crops make from the soil and the irrigation needs are determined mainly by the moisture content and by the matric potential of the soil. Soil moisture is defined as the amount of water contained in a given volume of soil and allows to calculate the sheet of water stored in the soil, while the matric potential of the soil is a measure of the force or suction with which the soil holds water. This suction is composed by intermolecular forces of adhesion between water and the aggregates and soil particles and by attraction forces between the water molecules. Because it is a suction force, it is expressed in negative terms and determines the availability of water for the crop [3].

Therefore, for irrigation control the matric potential, soil moisture or both variables can be measured. In recent years, multiple sensors and systems have been developed to measure these variables in real time, either with manual or automated measurement [4]. To monitor the water content of the soil, moisture sensors that are based on the principles of time domain reflectometry TDR [5, 6], of capacitance and of apparent electrical conductivity among others are used. While for monitoring the matric potential of the soil different tensiometers have been developed, gypsum block type sensors covered with a granular matrix, thermocouple-type thermal sensors [7–10].

These sensors have been used for decades in multiple crops with different irrigation systems to apply irrigation at the right time, to determine the required amount of water and to evaluate the quality of the work. Since 1951 there have been reports of the use of electric resistance blocks to control irrigation in sugarcane crops in Hawaii and actually in many regions of the world, both humidity sensors and matric potentials continue to be used [11–17].

The sensors are installed in the area of greater development of roots in the crop to determine the right time to begin and to end the irrigation based on the previously established criteria, such as the matric potential range or optimum moisture content for the development of the crop without exposing it to water stress conditions either by excess or deficit and in this way avoiding the reduction of crop productivity and water losses due to excessive application of irrigation [18, 19].

The matric potential of the soil is measured by means of a system composed by sensors that are inserted in the soil in the area with the greatest presence of roots in the crop (25 cm to 40 cm deep), by a system of excitation and data acquisition of the sensors, by a component of data transmission and an application that allows access to information in real time. In the market there is a great offer of sensors and equipment used for this purpose, but generally these are expensive [20, 21]. They are not calibrated for specific conditions, nor do they have the criteria that must be taken into account for each crop in their different stages of growth and development. This generates limitations in the assimilation of these technologies and therefore research must be done to know the soil-water-plant interaction, develop sensors and measurement devices of low cost, high precision, easy to use and that allow producers access technology for water management in different crops and in this way contribute to the efficient management of soil and water resources [4].

In the agroindustrial sector of sugarcane in Colombia, a system for monitoring the soil water status has been implemented. It consists of an IoT (Internet of Things) network that works through stations with sensors of matric potential of the soil in a set of nodes that are integrated into a hub. This is responsible for acquiring the data locally and transmitting them bi-directionally and wirelessly between stations and the Cenicaña server, to view them on mobile devices.

The Internet of Things or IoT is known as the ability to create connections between everyday objects, processes, internet and people. It thus achieves an interconnection in everyday life, sharing information and storing it in databases properly created for that purpose [20], a principle that was applied in the proposed monitoring system [21]. The IoT network generated in Cenicaña for monitoring the matric potential of the soil is integrated into the development that the sector has been innovating, such as precision agriculture, since its implementation frames the use of intelligent irrigation techniques that contribute to improving efficiency in the use of water and the profitability of crops [21].

This work was developed with the purpose of evaluating the use, functioning and performance of a monitoring network of matric potential of the soil, for the programming of irrigation in sugar cane crops in the Cauca river valley, Colombia.

2   Materials and Methods

The research was developed in several stages. The first one consisted in the definition of the variable or variables to be measured, based on criteria such as ease of measurement, precision, availability of sensors for measurement, costs, automated measurements, among others. The second one consisted in the evaluation of different sensors to measure the selected variable. The third one consisted in the definition of technical criteria (ranges of values to start and to end irrigation) to use the sensors in the programming of irrigation in the sugarcane crops. The fourth stage consisted in the start-up of a network or monitoring system.

This document presents the procedure and some results obtained in the fourth stage of the investigation, which consists of the implementation of the monitoring system.

2.1   Components of the Matric Potential of the Soil Monitoring System

2.1.1   Matric Potential Sensor

The sensors used for the measurement of the matric potential of the soil consist of two electrodes embedded in a porous cylinder wrapped in a metal mesh that allow the entry and exit of water from the sensor. The operation of the matric potential sensor is based on the principle of the variable electrical resistance. After inserting the sensor into the soil, it is balanced with the humidity content of it when the soil is wet and also the sensor. So the current conducts very well, but if the sensor is dry it resists the passage of the current. In this way, the measurement of the current flow through the sensor is used to calculate the matric potential of the soil by means of a calibration function. The matric potential sensors of this type need to be calibrated for each soil where they are installed.

2.1.1.1. Calibration of the Matric Potential Sensors

Since the sensors used to determine the matric potential of the soil do not directly measure this variable, it is necessary to calibrate it for each particular soil, this when it is necessary to measure the matric potential accurately, as it was in our case. Otherwise, a generic calibration for the sensor can be used.

In the place where the sensors will be installed, a soil sample is taken (at the depth where the sensors will be installed) of approximately 500 grams, trying to cover the sensor's exploration area, that is, in an area of 10 cm. depth and approximately 10 cm in diameter. The soil sample is packed in a plastic bag, taken to the soil laboratory and dries it in the air.

Once the soil is dry, it is passed through a No. 4 sieve (4.75 mm mesh opening), then the soil is packed in a 6 cm diameter and 8 cm high cylinder and the sensor of matric potential is inserted into this cylinder, then it is put in saturation by capillarity from 24 to 36 h. When the soil sample has been saturated with the sensor, it is exposed to different potentials in the Richards pressure cooker and at each potential the sensor output value (kOhms) is measured. In parallel, the temperature of the sample is measured to adjust the output signal of the sensor. Replications of three or more samples are made for each calibration. Once the values of both the matric potential and the sensor output have been obtained, the calibration curve is constructed (Fig. 6) and a polynomial is generated.

2.1.2 Cenilogger

The Cenilogger is an electronic circuit designed in Cenicaña for reading sensors and data storage. In a first version, it was based on the Arduino platform [19, 20, 22, 23], as it is easy to use free hardware and software, accessible and with different microcontrollers according to the requirements of the application. This development platform also has a simple and flexible programming environment [24].

The Cenilogger has been developed in Cenicaña and currently has version II. It is a modular connection model that uses low cost and easily accessible components available on the market, designed to read sensors or measuring devices such as rain gauges, anemometers, temperature sensors of different types, environmental humidity sensors, voltage, current, among others. Its modules interact with each other to make sensor readings, store in a microSD memory the data records obtained and transmit them in serial mode using an ATMega 2560 as controller. Diagram in Fig. 1.

A resistive transducer was generated for the construction of the specific Cenilogger to read the matric potential sensors. Two types of readers were also designed and built, one to be used manually and to read the sensors one by one with physical data recording in formats previously designed for that purpose. Another to automated reading that allows the data to be recorded, in a continuous and programmed way, through time to read six matric potential sensors and a soil temperature sensor, (Fig. 3), in addition to having a data transmission module [25]. Figure 2 shows the two types of datalogger developed.

The Cenilogger uses a transducer circuit whose main function is to include this variable resistance in an electronic circuit to obtain the voltage according to the changes in the moisture content of the soil. Another function of the transducer is to

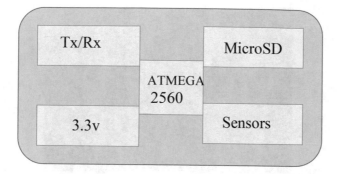

Fig. 1. Modular diagram of the Cenilogger

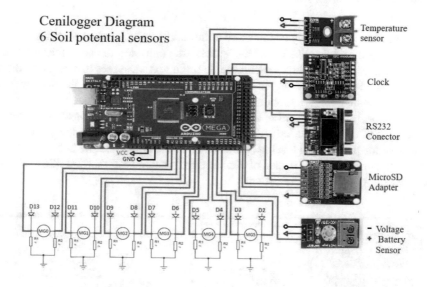

Fig. 2. Cenilogger diagram adapted to the matric potential of the soil

avoid that a galvanic corrosion effect occurs in the sensor due to the constant flow of current through it, which causes faults in the measurements made with the sensor [18]. Figure 3 shows the two types of datalogger developed.

2.2 IoT Network Applied to the Sugarcane Crops Using Cenilogger to Monitor the Matric Potential of the Soil

The IoT network of Cenicaña is a hybrid network based on ZigBee and mobile wireless technologies (GPRS, HSDPA, UMTS) that integrate acquisition, processing and communication [21], functions implemented through XBee® radio frequency modules and cellular modems.

Fig. 3. Cenilogger for measuring sensors manually (a) and Cenilogger for automated measurements of up to six sensors (b).

A station is composed by one or more nodes and a hub. A hub is the set between an XBee® receiver and a cellular modem, which are connected to each other by RS232. They configure a wireless link to send the data via text messages to a receiver modem [26]. The data is stored both in the datalogger and in a server, after the data arrives at the server is managed by a mobile application.

Each node is composed of a Cenilogger to which 6 matric potential sensors are connected, in addition to an XBee® module. The radiofrequency modules are responsible for providing a local network between the nodes and the hub of each place forming the monitoring stations of the matric potential of the soil [26]. Figure 4 illustrates the typology of the network.

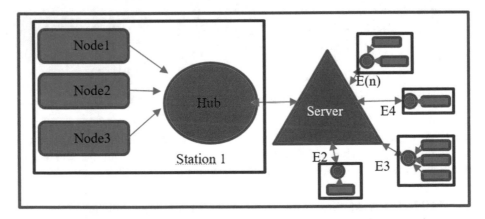

Fig. 4. Matric potential monitoring Cenicaña's network of the soil diagram.

Cenicaña uses a hybrid network configuration to increase the advantages that teams have separately, joining the DigiMesh™ typology (ZigBee Network) combined with a cellular point (Cellular Mobile Network) as shown in Fig. 5.

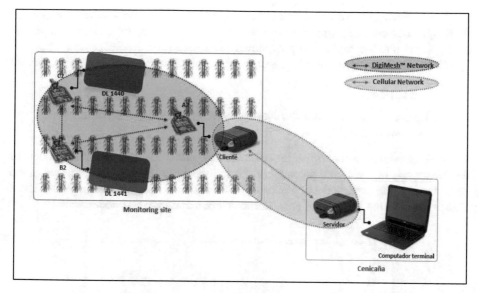

Fig. 5. Hybrid network deployed between the nodes of each station and the final modem towards the Cenicaña server.

3 Results

Figure 6 shows a typical calibration curve for a matric potential sensor in a soil classified as *Pachic Haplustoll*, a fine franc textural family.

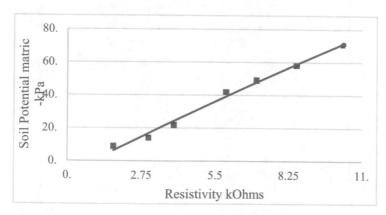

Fig. 6. Calibration curve for a matric potential sensor in a Pachic haplustoll soil, fine franc textural family.

The calibration curve was adjusted to a linear model, resistivity values are affected by temperature changes, which is why they must be corrected by a coefficient that is incorporated into the calibration polynomial.

The calibration curves must be generated for each of the soils where the sensors are installed.

3.1   Implementation of the Pilot Network for Monitoring the Matric Potential of the Soil

Currently, a pilot network with eight automatic stations consisting of thirteen nodes for the monitoring of the matric potential of the soil has been implemented, located in the central area of the department of Valle del Cauca and north of the department of Cauca. Figure 7 shows the location of each of the stations and installed nodes.

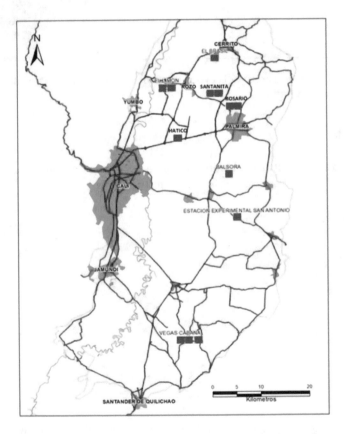

Fig. 7. Location map of the matric potential monitoring stations

3.1.1   Operation and Performance of Monitoring Stations

The network of pilot monitoring stations for irrigation administration was evaluated as part of sugarcane crops in open weather conditions.

The problems experienced in the field have been of a diverse nature, both electrical and because of environmental conditions, failures of the communication system, and others implicit in the manipulation of the equipment. One of these problems was the underestimated initial calculation of the energy system, for which there were repetitive blackouts in the stations.

All difficulties have been rigorously evaluated in order to find solutions, some have been restructuring the prototype of the Cenilogger, others have led to improve the gain of communication antennas and/or change the location of the hubs. Other solutions have led to the standardization of protocols such as maintenance in the field, construction of Cenilogger and its validation in the laboratory.

3.2 Programming the Risks in the Cane Crops from the Information of Matric Potential of the Soil Obtained in the Monitoring Stations

The matric potential of the soil data recorded in each of the monitoring stations was used to program the irrigation in each of the sugarcane fields where the station was installed.

Figure 8 shows the dynamics of the matric potential of the soil measured in a cane crops cycle at two depths, 30 cm and 45 cm. The records obtained with the sensors installed at 30 cm are used to program the irrigation based on the aforementioned criteria for the sugarcane crops. The records obtained with the sensors installed at 45 cm depth are used to verify that the water applied by irrigation reached that depth, this as an indicator of the quality of irrigation.

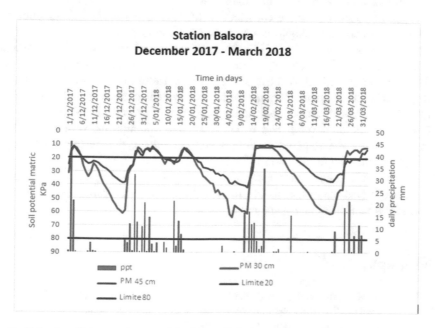

Fig. 8. Behavior of the matric potential of the soil during a cane production cycle (13 months), in a soil with a fine franc texture.

In this figure we can observe the changes in the matric potential of the soil due to the contributions of water by precipitation and irrigation (rising peaks) and to the water outlet by the extraction of the crop (evapotranspiration). After a rainfall or an irrigation event, part of the water infiltrates and accumulates in the soil profile, which increases the availability of water for the crop. This process is reflected in the sensor and generates a change in the output signal, for example when going from a value of −60 kPa to −10 kPa. On the contrary, as the crop extracts water from the soil due to evapotranspiration, it is also reflected in the sensor when passing high (less negative) values, for example −10 kPa, at low values (more negative) −70 kPa, which indicates the gradual reduction of water availability in the soil for plants. Once the sensors reach average values of −80 kPa ± 5 kPa, it indicates the right time for the beginning of the irrigation and average values of −20 kPa ± 5 kPa the time to finish the irrigation.

The matric potential data is shown to the users through a curve similar to the one presented in Fig. 8, which can be visualized by a mobile application. This tool allows timely decision making for irrigation management in cane crops in the Cauca River Valley, Colombia.

4   Conclusions and Recommendations

- The sensors used to monitor the matric potential of the soil showed good performance in the range of −30 kPa to −150 kPa. This makes them suitable for use in irrigation programming with the different application methods commonly used in the sugarcane crops in the Cauca river valley.
- The use of Arduinos for the construction of dataloggers is limited for pilot measurement systems and should not be used on a commercial scale.
- The system of transmission of data through the use of cellular networks presents difficulties due to low coverage, mainly in rural areas where sugarcane is grown.
- The monitoring method used has proven to be a useful tool for users to make right decisions based on quantitative measurements, which contributes to the efficient use of soil and water resources and the improvement of crop profitability.
- It is necessary to develop and evaluate a robust monitoring system that works properly under the required environmental conditions, that is low cost, easy to install and that the information is easily understood by users.
- Given the limitations in current communication methods and systems it is necessary to develop local data transmission networks that allow the capture of information and access to it from any place at very low cost.

Bibliography

1. Bittelli, M.: Measuring soil water content: a review. Horttechnology 21(3), 293–300 (2011)
2. Kalita, H., Palaparthy, V.S., Baghini, M.S., Aslam, M.: Graphene quantum dot soil moisture sensor. Sens. Actuators B Chem. 233, 582–590 (2016)
3. Hillel, D.: Enviromental Soil Physics, p. 771. Acad. Press USES (1998)

4. Hincapié, E.: Claves para utilizar sensores en labores de riego. Carta Informativa, Centro de Investigación de la Caña de Azúcar **2**(5), 6–7 (2017)
5. Topp, G.C., Davis, J.L., Annan, A.P.: Electromagnetic determination os soil-water content: Measurement in coaxial transmissions lines. Water Resour. Res. **16**, 574–582 (1980)
6. Evett, S.R.: The TACQ computer program for automatic time domain reflectometry measurements: I. Design and operating characteristics. Trans. ASAE **43**(6), 1939–1946 (2000)
7. McCarthy, A.C., Hancock, N.H., Raine, S.R.: Development and simulation of sensor-based irrigation control strategies for cotton using the VARIwise simulation framework. Comput. Electron. Agric. **101**, 148–162 (2014)
8. Nolz, R., Kammerer, G., Cepuder, P.: Calibrating soil water potential sensors integrated into a wireless monitoring Network. Agric. Water Manag. **116**, 12–20 (2013)
9. Pedro, C.M., Gimenez, A., Porto, L.F., Porto, L.H.: Principles and applications of a new class of soil water matric potential sensors: the dihedral tensiometer. Procedia Environ. Sci. **19**, 484–493 (2013)
10. Evett, S.R., Schwartz, R.C., Casanova, J.J., Heng, L.K.: Soil water sensing for water balance. ET and WUE. Agric. Water Manag. **104**, 1–9 (2012)
11. Paraskevopoulos, A.L., Singels, A.: Integrating soil water monitoring technology and weather based crop modelling to provide improved decision support for sugarcane irrigation management. Comput. Electron. Agric. **105**, 44–53 (2014)
12. Wiedenfeld, B.: Scheduling water application on drip irrigated sugarcane. Agric. Water Manag. **64**, 169–181 (2004)
13. Varble, J.L., Chávez, J.L.: Performance evaluation and calibration of soil water content and potential sensors for agricultural soils in eastern Colorado. Agric. Water Manag. **101**, 93–106 (2011)
14. Blonquist, J.M., Jones, S.B., Robinson, D.A.: Precise irrigation scheduling for turfgrass using a subsurface electromagnetic soil moisture sensor. Agric. Water Manag. **84**, 153–165 (2006)
15. Cardenas-Lailhacar, B., Dukes, M.D.: Precision of soil moisture sensor irrigation controllers under field conditions. Agric. Water Manag. **97**, 666–672 (2010)
16. Moreno, L.P.: Respuesta de las plantas al estrés por déficit hídrico. Una revisión. Agron. Colomb. **27**(2), 179–191 (2009)
17. Flores, W., Estrada, H., Jiménez, J., Pinzón, L.: Effect of water stress on growth and water use efficiency of tree seedlings of three deciduous species. Terra Latinoam. **30**(4), 343–353 (2012)
18. Valente, A., Morais, R., Couto, C., Correia, J.H.: Modeling, simulation and testing of a silicon soil moisture sensor based on the dual-probe heat-pulse method. Sens. Actuators A Phys. **115**(2–3), 434–439 (2004)
19. Ruiz Canales, A., Oates, M.J., Pérez Solano, J.J., Molina Martínez, J.M.: Sensores de bajo coste aplicados al control de los cultivos. interempresas.net (2018). https://www.interempresas.net/Horticola/Articulos/208300-Sensores-de-bajo-coste-aplicados-al-control-de-los-cultivos.html. Accessed 03 July 2018
20. Ivan-rios, J., Castro-silva, J.A.: Sistema de Riego Basado En La Internet De Las Cosas (IoT), no. November 2017
21. Karim, F., Karim, F., Frihida, A.: Monitoring system using web of things in precision agriculture. In: 12th International Conference on Future Networks and Communications, vol. 110, pp. 402–409 (2017)

Author Index

A

Abad, Juan, 202
Afroz, Farhana, 136
Akhter, Ayeasha, 136
Alarcón, Claudio, 89
Ali, Abid, 146
Álvarez-Castillo, Paola, 218
Alzate Velásquez, Diego Fernando, 106
Anannya, Tasmiah Tamzid, 136, 146
Arias-Carrera, Rosa, 218
Aziz, Faisal Ibn, 146

B

Bashar, Saad Bin, 136
Bayzed, Moin, 146
Borja-Goyes, Oswaldo, 73, 218

C

Cajas-Morales, Leidy, 218
Caquilpan, Victor, 89
Carbonell González, Javier Alí, 249
Cardenas, Pedro Fabian, 1, 21
Cárdenas, Roberto, 89
Carrera-Villacrés, David, 73, 218
Cau, Pierluigi, 162
Chasi, Paúl, 202
Clotet, Roger, 42, 202

D

De Giudici, Giovanni, 162
Dewan, Tanu, 136
Duarte, Freddy, 234

E

En-Nejjary, Driss, 190
Enríquez-Herrera, Karla, 218
Espina, Enrique, 89

F

Farez, Juan, 202

G

García Cárdenas, Diego Alejandro, 106
García-Cedeño, Andrea, 42, 202
González-Farinango, Tania, 218
González-Riera, David, 73, 218
Guamán-Pineda, Erika, 218
Guerrón-Varela, Edgar, 73
Guillermo, Juan Carlos, 42, 202
Guitiérrez-Puetate, Solange, 218

H

Hernández, Giovanny, 21
Hernández, Roberto, 89
Herrera, Eric F., 1
Hincapié Gómez, Edgar, 249
Huerta, Mónica, 42, 202
Huircán, Juan, 89

I

Islam, Muhammad Nazrul, 146

© Springer Nature Switzerland AG 2019
J. C. Corrales et al. (Eds.): AACC 2018, AISC 893, pp. 261–262, 2019.
https://doi.org/10.1007/978-3-030-04447-3

J
Jahan, Mahmuda Rawnak, 146
Jannat, Fatima, 136
Jimenez, Andres Fernando, 1, 21

K
Kang, Myoung-Ah, 190

L
Lamadrid, Alfonso P., 21
Ledezma, Agapito, 58
Llano, Gonzalo, 120
López, Yaneth Patricia, 58

M
Mithra, V. S. Santhosh, 177
Montalvo-Alvarado, Paula, 73, 218
Morales, Raúl, 89
Moreno-Chauca, Andrés, 73, 218
Muñoz, Carlos, 89

O
Orozco, Oscar A., 120
Ortiz, Brenda V., 1, 21

P
Painemal, Necul, 89
Palacios, Jordi Rafael, 106, 234
Pinet, François, 190

R
Rabbi, Md. Fazle, 146
Ramón Valencia, Jacipt Alexander, 106
Rivas-Lalaleo, David, 42
Rivera, Wilfred Fabian, 58
Roje, Tomislav, 89
Rony, Shamsuzzaman, 146
Rubio-Gallardo, Richard, 218
Ruiz, Antonio, 1

S
Sáez, Doris, 89
Sánchez Benítez, Juliana, 249
Sandoval-Plaza, Tatiana, 218
Santos, Germán Ricardo, 234
Suárez, Luis Javier, 58

T
Tarik, Ismat, 136

U
Unda-León, Nelson, 218

V
Vargas, Carolina, 89
Velarde-Salazar, Paul, 218

Y
Yeazdani, Anika, 146

Printed in the United States
By Bookmasters